Luca Bortolussi

Computational Systems Biology with Constraints

Luca Bortolussi

Computational Systems Biology with Constraints

Stochastic Modeling of Biological Systems with Concurrent Constraint Programming

VDM Verlag Dr. Müller

Impressum/Imprint (nur für Deutschland/ only for Germany)

Bibliografische Information der Deutschen Nationalbibliothek: Die Deutsche Nationalbibliothek verzeichnet diese Publikation in der Deutschen Nationalbibliografie; detaillierte bibliografische Daten sind im Internet über http://dnb.d-nb.de abrufbar.

Alle in diesem Buch genannten Marken und Produktnamen unterliegen warenzeichen-, marken- oder patentrechtlichem Schutz bzw. sind Warenzeichen oder eingetragene Warenzeichen der jeweiligen Inhaber. Die Wiedergabe von Marken, Produktnamen, Gebrauchsnamen, Handelsnamen, Warenbezeichnungen u.s.w. in diesem Werk berechtigt auch ohne besondere Kennzeichnung nicht zu der Annahme, dass solche Namen im Sinne der Warenzeichen- und Markenschutzgesetzgebung als frei zu betrachten wären und daher von jedermann benutzt werden dürften.

Coverbild: www.purestockx.com

Verlag: VDM Verlag Dr. Müller Aktiengesellschaft & Co. KG
Dudweiler Landstr. 125 a, 66123 Saarbrücken, Deutschland
Telefon +49 681 9100-698, Telefax +49 681 9100-988, Email: info@vdm-verlag.de
Zugl.: Udine, University of Udine, PhD Thesis, 2007

Herstellung in Deutschland:
Schaltungsdienst Lange o.H.G., Zehrensdorfer Str. 11, D-12277 Berlin
Books on Demand GmbH, Gutenbergring 53, D-22848 Norderstedt
Reha GmbH, Dudweiler Landstr. 99, D- 66123 Saarbrücken
ISBN: 978-3-639-08875-5

Imprint (only for USA, GB)

Bibliographic information published by the Deutsche Nationalbibliothek: The Deutsche Nationalbibliothek lists this publication in the Deutsche Nationalbibliografie; detailed bibliographic data are available in the Internet at http://dnb.d-nb.de.

Any brand names and product names mentioned in this book are subject to trademark, brand or patent protection and are trademarks or registered trademarks of their respective holders. The use of brand names, product names, common names, trade names, product descriptions etc. even without
a particular marking in this works is in no way to be construed to mean that such names may be regarded as unrestricted in respect of trademark and brand protection legislation and could thus be used by anyone.

Cover image: www.purestockx.com

Publisher:
VDM Verlag Dr. Müller Aktiengesellschaft & Co. KG
Dudweiler Landstr. 125 a, 66123 Saarbrücken, Germany
Phone +49 681 9100-698, Fax +49 681 9100-988, Email: info@vdm-verlag.de

Copyright © 2008 VDM Verlag Dr. Müller Aktiengesellschaft & Co. KG and licensors
All rights reserved. Saarbrücken 2008

Produced in USA and UK by:
Lightning Source Inc., 1246 Heil Quaker Blvd., La Vergne, TN 37086, USA
Lightning Source UK Ltd., Chapter House, Pitfield, Kiln Farm, Milton Keynes, MK11 3LW, GB
BookSurge, 7290 B. Investment Drive, North Charleston, SC 29418, USA
ISBN: 978-3-639-08875-5

To Carla

List of Figures

List of Tables

Introduction

Πάντα ρει - *Everything Flows*

Eraclito

In the last thirty years, the constant improvement of experimental techniques in molecular biology led to a massive increase in data available to scientists. Consequently, there has been a growth also in our comprehension of the basic mechanisms of living beings.

This increment of biological understanding, nevertheless, happened mostly in a *reductionist perspective*: living beings are studied at the cellular level by analyzing in deep detail each single molecule involved in biological processes. The resulting picture is that of a complex network of interactions between bio-molecules. The properties of such networks, however, are not reducible to their components, but rather they are a feature of the interaction patterns. This is the point where the reductionist approach shows its limitations: even if we had a complete map of interactions of each single molecule involved in cells, still we could not understand the functioning of life. This is the main propulsor of a shift of paradigmatic perspective we are observing nowadays: from reductionism to a systemic approach. In the systemic point of view, the focus is on the interaction between biological entities, in order to understand the properties of the whole system, not reducible to single parts. This change of focus is so deep that a new science is emerging from it: systems biology [142].

At the heart of this new discipline there is the need of formal tools for describing biological networks, building mathematical models and analyzing them in order to understand the functioning of such systems. Moreover, their dimension is so big that it is impossible to model them by hands, hence the necessity of computational tools both in the modeling phase and in the analysis one.

In this scenario we are assisting to the convergence of techniques coming from different areas: theoretical and experimental biology, mathematics (biomathematics, theory of dynamical systems), and computer science (stochastic simulations, formal theory of concurrence). In particular, mathematics and computer science are starting an integration process with biology to devise adequate formal tools, resulting in a moment of great scientific ferment [142, 2].

Systems biology has its own prophets [46]: back at the beginning of twentieth century, thinkers like D'Arcy Thompson [228] and Von Bertalanffy [237] pointed clearly at the need for a systemic point of view in the study of biological systems. Later, this necessity was underlined by the thought stream of cybernetics. Scientists like Wiener, Von Neumann, Shannon, Bateson, just to cite a few, set up the main principles in the functioning of complex systems, like feedback regulation mechanisms [238, 121]. Despite these enlightened minds, the systemic approach never imposed itself in the mainstream biology. This may have been caused by a lack of experimental data, by a lack of existing knowledge or

11

by a lack of computational power capable of tackling real life systems. The impressive increase in biological knowledge and in computational technology makes time mature for this discipline to impose itself [2].

System biology needs both methods to describe efficiently biological systems and automatic techniques to analyze the obtained models. In this direction, the theory of dynamical systems offers a tool extremely well developed from a mathematical point of view: differential equations [83, 236]. However, modeling by differential equations is a complex activity that requires a substantial experience, especially when the dimension of systems under study gets big.

On the other hand, computer science faces this area equipped with a set of techniques developed to analyze another class of complex systems, presenting several analogies with biological ones: concurrent software systems, composed by many processes communicating among them. In particular, computer scientists developed in the last twenty years both languages capable of describing in a simple and effective way the main features of these systems and analysis techniques calibrated on these formalisms [130, 123, 172, 174, 92, 13]. The main responsible of the dynamics of software systems is the communication among different processes (or agents): this simple form of interaction is able to produce (and describe) the huge variety of observable behaviours exhibited [171, 173]. In this light, what we observe in a software system is an emerging behaviour deriving from the interaction through communication among the different processes.

The parallelism with biological systems is straightforward: abstracting from biochemical details of molecular interaction, we can describe these systems as a network of interacting entities. Substituting the term "communicating processes" to the term "interacting agents", it is clear why techniques originally developed for software systems can be also used for biological ones [203].

Probably *process algebras* are the most widespread languages used to describe software systems at the abstraction level of communication. Their stochastic variants, developed in computer science as tools for analyzing performances [125], have been used also to model biological systems, with interesting perspectives [197]. There are also other formalisms used in biological modeling that have an origin similar to process algebras, like stochastic Petri nets [190, 117, 15, 110, 239].

However, *a cell is not a software program*: biological systems have a set of specific features that cannot be ignored in the modeling phase (like the high compartmentalization, the different time scales involved, etc.). It is therefore necessary to adapt these formalisms to the biological reality, tuning consequently the analysis techniques.

Among stochastic process algebras (from now on, SPA), the ones that have been mostly used in biological modeling are stochastic π-calculus [194, 195] and PEPA (Performance Evaluation Process Algebra) [128, 44]. Despite their differences, they are based on the simple primitive action of synchronization between two (π-calculus) or more (PEPA) processes. In addition to synchronizing, processes localize some interactions and perform a *don't care* choice among several alternative behaviours. These simple agents are then composed together in parallel, using a dedicated operator. This ingredient makes these languages compositional: the user can model the single parts separately and then glue the models together essentially by syntactic juxtaposition. This feature is perfectly fitted for the type of knowledge we have of biological systems: we know the interaction of each single piece and we want to understand the dynamics of the whole. Therefore, compositionality seems a crucial feature that a modeling language should possess. This is probably the main limitation of other formalisms coming from computer science, like stochastic Petri nets.

The models constructed using process algebras are discrete stochastic processes evolving in continuous time, namely Continuous Time Markov Chains [181]. Incidentally, this is the same kind of stochastic process arising in direct simulation of biochemical reactions, based on arguments coming from physical considerations [101]. It's remarkable that the algorithmic engine used in computational chemistry to perform computer simulations is the same used to simulate process algebra models, i.e. Gillespie algorithm [101, 102].

Stochastic process algebras like π-calculus and PEPA are simple and elegant, but they have some drawbacks. According to our judgement, among the principal limitations of SPA there are the inability of using classical computations in the models and the impossibility of expressing arbitrary relations between elements of the models. For example, the introduction of (qualitative or quantitative) spatial information is needed while describing systems where the relative orientation or the relative distance between elements is important. This is the case, for instance, of high detailed models of protein-protein interaction or of models describing also the (usually inhomogeneous) spatial distribution of molecules inside cells. When one has to model such systems, usual process algebras are of no help, and different discrete mathematical formalisms must be used. On the contrary, differential equations have a more general range of applicability, though the assumption of continuity and determinism cannot always be made safely for the biological processes under consideration. We deem that it would be interesting to have a process algebra formalism that has the possibility of encoding in a simple way different kinds of information, in order to be used as a general purpose language for modeling systems at different levels of detail, similarly to differential equations. Essentially, one would like to have a simple basic language that can be extended by a programming activity of the user.

The previous statement can be seen as a guiding line for this work, done by the author during its PhD [26]. Its topic is, in fact, the identification of a candidate language possessing the desired form of extensibility. The features of the language are then studied focusing on their repercussions on the modeling activity.

The key instrument we used to give flexibility to the language are *constraints*. Constraints are relations restricting the values that the variables of the system can take. Essentially, they are representable as first-order logical predicates interpreted on a predefined model. In computer science, the idea of constraint gave rise to different classes of languages, like *constraint logic programming* [10], whose aim is searching good solutions in a state space, and *concurrent constraint programming* (CCP [214]), a process algebra devoted to describing concurrent processes. CCP-agents compute by adding constraints (**tell**) into a "container" (the *constraint store*) and checking if certain relations are entailed by the current configuration of the constraint store (**ask**). The communication mechanism among agents is therefore asynchronous, as information is exchanged through global variables. In addition to **ask** and **tell**, the language has all the basic constructs of process algebras: non-deterministic choice, parallel composition, procedure call, and the declaration of local variables. This dichotomy between agents and the constraint store can be seen as a form of separation between computing capabilities (pertaining to the constraint store) and the logic of interactions (pertaining to the agents). This feature makes it an ideal candidate for our purposes: an user that wishes to encode specific information in the model has just to code it using suitable constraints, with no need to modify the machinery of the language. From a general point of view, the main difference between CCP

13

and other process algebras like π-calculus resides really in the computational expressivity of the former. π-calculus, in fact, has to describe everything in terms of communications only, a fact that may result in cumbersome programs in all those situations in which "classical" computations are directly or indirectly involved.

In order to use CCP for biological modeling, we had to define a stochastic variant of it [25]. This is obtained by adding an exponentially distributed stochastic duration to all the instructions interacting with the store, a standard praxis in stochastic languages. What is non-standard is the fact that these rates are not real numbers (like in standard process algebras) but rather functions mapping the current configuration of the store into a real number. This makes duration of processes explicitly dependent on context. The language is then equipped with two transition relations: an instantaneous and a stochastic one. The instantaneous transition is made finite and confluent by imposing suitable syntactic conditions, hence the evolution of the system is given in terms of the stochastic relation. The underlying stochastic model can be both a Continuous Time Markov Chain or a Discrete Time Markov Chain [181], in order to take into account continuous time only when necessary. We also designed an interpreter for the language in Prolog, exploiting the constraint logic programming libraries of SICStus Prolog [93] in order to manage the constraint store. The interpreter's engine is based on the famous Gillespie's algorithm, and it allows to perform stochastic simulations of the models.

With the language at our disposal, next step was to investigate its usage as a modeling tool for biological systems [36]. The main idea while building a model with a process algebra like stochastic π-calculus is to identify each entity of the system with a process and each reaction (or, more generally, interaction) with a communication on a dedicated channel [197]. The number of entities of the same type has a direct counterpart in the number of instances in parallel of the corresponding agent. In sCCP, instead, we have more freedom in choosing the point of view, and we can adapt it to the specific domain of the model. For instance, while dealing with biochemical reactions, our approach is *reaction-centric*. The different entities of the system are associated to variables of a special kind, called stream variables, representing quantities varying over time. On the other hand, the different interactions are encoded as processes, acting by modifying some variables of the system. Within this scheme, we are able to find a straightforward mapping of biochemical reactions (described by means of chemical arrows). Gene regulatory networks, instead, can be dealt by mixing the *reaction-centric* style with an *entity-centric* one: agents correspond to genes and to reactions involving only proteins. The presence of functional rates allows to encode different forms of chemical kinetics, like Michaelis-Menten or Hill kinetics [83], describing the speed of reactions at different levels of abstraction. This is a distinguishing feature w.r.t. standard SPA, where only the mass action kinetics can be used, giving more flexibility to sCCP. Indeed, the importance of functional rates has been recently recognized also by other practitioner of biological modeling with SPA: an example is bioPEPA [53]. The use of more complex kinetics, from a mathematical point of view, can be seen as a form of simplification of the underlying stochastic model, replacing set of states with single states without strong alterations of the probability of traces.

In order to facilitate the modeling phase, we designed some simple libraries of processes, that can be used as basic building blocks to be composed together. This is the first fundamental step to design a graphical interface for model building, which can be used also by biologists.

Once a model has been written, we can feed it to the interpreter, perform some stochas-

14

tic simulations, and study its dynamics. However, SPA are equipped with a set of tools that can be used for analyzing models, like stochastic model checking [211]. One of the most well-developed model checkers for stochastic systems is PRISM [155], which allows to verify if a CTMC satisfies properties specified with formulae of Stochastic Continuous Logic [11], a temporal modal logic enriched with operators expressing probabilistic information. In order to use such tools, we defined a translation procedure associating to each sCCP program a (stochastically) equivalent one written in the language of the PRISM model checker.

An important part of protein interaction in cells consists of the formation of big molecular complexes, constituted by several different proteins stuck together by chemical bonds. These complexes are constructed incrementally, hence a big number of different intermediate sub-complexes can be present at the same time in a cell. The combinatorics of these sub-complexes can be disarming, and some methods are available to provide compact descriptions. We recall the graphical language developed by Kohn, Molecular Interaction Maps [147], and κ-calculus [65, 64], a rule-based process algebra capable of defining complexes implicitly, using dynamical rules independent from the context. To show the flexibility of sCCP, we defined a simple encoding both of κ-calculus and of MIMs, making use again of a mixture between *reaction-centric* and *entity-centric* modeling styles.

One of the hypothesis made while modeling with SPA is the spatial homogeneity of systems under consideration [101]. This hypothesis is questionable even for biochemical reactions in vivo, and it is a complete non-sense for other biological processes, where the spatial information becomes an essential ingredient. One of such process is the folding of a protein. A protein is a a polymeric chain built up from 20 different aminoacids, assuming a peculiar 3D conformation (native structure) that determines its function. The process of reaching this configuration is called folding, while the problem of predicting the native structure given only the sequence of composing amino acids is called protein structure prediction [179].

In sCCP, we can use constraints to encode spatial information, like cartesian coordinates. Therefore, we can model the folding of a protein, describing all the entities (atoms or aminoacids, depending on the granularity of the description) and forces (seen as stochastic interactions) into play. However, we can also take a different point of view, designing a concurrent algorithm to tackle the protein structure prediction problem, which can be modeled as the search of the configuration of minimum free energy. Unfortunately, even very coarse abstractions of it are known to be NP-hard [61]. To worsen the situation, the energy functions to be minimized present an exponential number of local minima (cf. [179]).

The most striking feature, however, is not its mathematical difficulty, but the fact that Nature is able to fold correctly a protein in a very short time (milliseconds), "exploring" a very small portion of the search space, even if the main forces behind the process (i.e. protein-solvent interaction) have a "stochastic" nature (in the sense of statistical mechanics). In addition, the native configuration is shaped simply by the interaction among the atoms constituting the protein. In some sense, it is the concurrent interaction between these "simple" agents, that obey "simple" rules (the laws of physics), that determines both the protein's native structure and the dynamics of the process for reaching it. Therefore, one of the key ingredient that enables Nature to be so efficient can be argued to be this total concurrency itself (with fast communication channels).

The previous reflections are the starting points of the attempts we made so far in modeling the protein structure prediction problem in a concurrent setting. In [28, 29], we associated an independent process to each aminoacid, modeled here in a simplified way as a single center of interaction (a common approach, cf. [223]). These agents interact by exchanging information about their spatial position, using this knowledge to move in the space, trying to reach the configuration of minimum free energy. The moving strategy used is a Monte-Carlo one: moves lowering the energy are always performed, while moves rising the energy are executed with a certain probability depending on the difference of potential. The whole simulation can be written in sCCP, exploiting the stochastic semantics (in this case, the one with discrete time) in order to model the probabilistic search of the state space. The results were encouraging, even if the coarseness of the energy model used forbade to obtain decent predictions for whole proteins.

To tackle these problems, in [30, 38] we embedded the previous framework in a multi-agent scheme [169] designed for optimization tasks, implementing the whole system both in sCCP and in C (an ad-hoc multi-threading version, in order to speed it up). In this new model we adopted a different potential, still representing the aminoacids as a single center of interaction and identifying them with a concurrent process. In addition, we introduced some higher level agents, which have the task of both coordinating the exploration of the state space performed by the aminoacid agents and introducing some form of cooperation among them. This cooperative action helps the aminoacids to form some local patterns, like helices and sheets, that are very common in proteins. The enhancements introduced here improve a lot the results obtained in terms of both the stability of the simulation and the quality of the predictions of the native structures.

There is another fascinating problem where sCCP enters in the picture, namely the relation between process algebras and differential equations. These two approaches are radically different: the models built using SPA are discrete and stochastic, and they are generally used to infer pattern of dynamical evolution of the system. On the other hand, ODEs provide a continuous and deterministic model and can be used to formulate quantitative predictions. Another fundamental difference between SPA and ODEs is the point of view taken while modeling a system. With process algebras, in fact, one has to describe the logical patterns of interaction between the different entities: a process is associated to each biological entity and the interaction capabilities are represented as communication links among processes. This approach is internal, in the sense that it describes the system with an inner perspective, modeling each single entity and composing them together to obtain the global model. On the contrary, ODEs take an external approach, in which what is actually modeled are the mutual relationships among variables expressing concentrations of substances. This corresponds to the point of view of an experimenter observing the temporal evolution of the quantities it is interested in, and relating such quantities through mathematical laws.

In [32], we explored the possibility of defining procedures capable of converting one formalism into the other. Clearly, here there are two directions: associating ODE systems to SPA programs and SPA programs to ODEs systems. In both cases the translation should preserve the behaviour of the system it applies to, where for behaviour we mean the time trace for ODEs and the average evolution for stochastic systems.

First of all, we extended to sCCP a method developed by Hillston [129] for PEPA models [128]. The class of differential equations obtainable from sCCP turns out to be broader than the class for PEPA. This is related to the fact that stochastic rates in sCCP are functions rather than numbers. These equations relate to the average value (over all possible traces) of variables under interest. Essentially, they are a first-order approximation (hence neglecting fluctuations) of the ODE describing the true average [27]. However, there are cases where the behavior demonstrated by stochastic fluctuations cannot be averaged away, thus a translation procedure from SPA to ODE's must take into account these effects, if it has to be behaviorally invariant with respect to the modeled system. A possible solution is to use information about the model to fix a format for the target ODEs, (that will belong to the class of S-Systems [236]), using a set of stochastic traces to fit the remaining parameters in order to reproduce the same behaviour.

The passage from ODE to SPA is also an interesting problem, and it can be seen as a way of deducing, starting from a set of observations, the logical structure of processes and interactions describing the systems, i.e. the fine topology of the biological network under exam.

In this direction, the use of sCCP becomes fundamental. In particular, having at our disposal non-constant rates allows to define a simple automatic translation from general ODEs to sCCP programs, that exhibit the same behaviors and are provably coherent from a quantitative point of view. This translation suggests that the extension of usual process algebras with the ingredient of functional rates is non-trivial.

The book is divided in two parts. The first one covers the preliminaries needed to read the second part, where the personal contributions are found. Regarding the first part, Chapter 1 introduces the basics of Concurrent Constraint Programming, Chapter 2 introduces some notion of stochastic processes, Chapter 3 presents briefly the state of the art in modeling biological processes, covering both SPA and ODE models, Chapter 4 introduces the problems of protein folding and protein structure prediction, and Chapter 5 gives a brief overview of optimization techniques. The second part, instead, is structured as follows: Chapter 6 presents the stochastic version of CCP, Chapter 7 covers the usage of sCCP in modeling biological systems, Chapter 8 focuses on the use of sCCP for the protein structure prediction, and finally Chapter 9 analyzes the relation between stochastic process algebras and differential equations.

Part I

Basic

Chapter 1

Concurrent Constraint Programming

Concurrent Constraint Programming (from now on, CCP) is a programming language for concurrent systems, combining a syntax in the style of process algebras with the computational power of constraints. Introduced by Saraswat in his PhD thesis [214], it is still now considered one of the most advanced languages in the family of Concurrent Logic Programming languages [68].

In CCP, the usual Von Neumann's store is replaced by a store that contains constraints on the variables into play. Computations evolve monotonically: constraints can be added in the store, but never removed. The idea behind this approach is to attach to variables not a single value, but rather an interval of possible values, which is refined as long as new information, in the form of constraints, is available. CCP is a language suitable for modeling purposes, as it separates the description of the logics of interactions, expressed in a process algebra fashion, from the description of the properties of the modeled objects, encoded as constraints in the constraint store. In a CCP program, different agents can run in parallel, and the communication between them is performed through shared variables. In particular, agents can either add (`tell`) a constraint in the store, or they can check (`ask`) if a particular formula is satisfied by the current configuration of the system.

CCP, since its appearance, has been extended in different directions: probabilistic [74], distributed [116, 97], hybrid [114], temporal [186], just to cite a few.

In this Chapter, we give a brief overview of the language, focusing mainly on the definition of constraint systems, syntax and operational semantics, and on some of its extensions. The chapter is organized as follows: constraint system are recalled in Section 1.1, the syntax of the language is shown in Section 1.2 while its operational semantics is presented in Section 1.3. In Section 1.4 we spend some words on the probabilistic version of CCP, while in Section 1.5 we review some proposals for distributing it. Finally, in Section 1.6, we show how time-dependent quantities can be modeled.

1.1 Constraint Systems

Computations in CCP are performed through a monotonic update of the constraint store, which is usually modeled as a constraint system. We follow here a well established approach (cf. [215, 213, 69]), representing a constraint system as a complete algebraic lattice.

In the following, we assume as known basic concepts in logic and in domain theory, like partial orders, lattices, and so on. The reader wishing to review such notions is referred to [82] for an introduction on mathematical logic and to [66] for an overview on lattices and domain theory.

Constraint systems are constructed starting from the primitive notion of *simple constraints*, or tokens, together with a *compact entailment relation* $\vdash$ defined on them. This relation is then extended on set of tokens, and constraints are defined as sets of tokens closed w.r.t. $\vdash$. Essentially, this corresponds to the definition of a system of partial information in the sense of Dana Scott [217]. Usually, such a constraint system is derived from a first-order language together with a fixed interpretation, where constraints are formulas of the language, and a constraint c entails a constraint d, $c \vdash d$, if every evaluation satisfying c satisfies also d. Clearly, the predicate $\vdash$ must be decidable.

Formally, a simple constraint system is a pair $(D, \vdash)$, where D is the set of simple constraints, $\mathcal{D} = \wp_f(D)$ is the set of finite subsets of D, and $\vdash \subseteq \mathcal{D} \times D$ is the entailment relation satisfying the following properties (U, V range on $\mathcal{D}$ and u, v, w range on D):

- $U \vdash u$ if $u \in U$

- $U \vdash w$ if $U \vdash v$ for all $v \in V$ and $V \vdash w$.

The relation $\vdash$ can be extended to $\mathcal{D} \times \mathcal{D}$ by letting $U \vdash V \Leftrightarrow \forall v \in V(U \vdash v)$. Then, we can define constraints as subsets of D closed by (finite) entailment: $c \subseteq D$ is a constraint if $U \in \mathcal{D}$, $U \subseteq c$, and $U \vdash u$ implies $u \in c$. The set of constraints defined in this way is denoted by $\mathcal{C}$. The closure of a set $U \subseteq D$ is denoted by $\widehat{U} \in \mathcal{C}$. Entailment relation can be extended to $\mathcal{C} \times \mathcal{C}$ by letting $c \vdash d$ iff $c \supset d$, thanks to the closure property of elements of $\mathcal{C}$. This definition suggests that $\mathcal{C}$ can be given a lattice structure by defining the partial order $\sqsubseteq$ as the inverse of the entailment relation, $c \sqsubseteq d \iff d \vdash c$, and the least upper bound operation as the closure of the union on the set of constraints: $c \sqcup d = \widehat{c \cup d}$. The bottom element is, therefore, $true = \emptyset$ and the top element is $false = D$. The resulting lattice can be shown to be algebraic and complete, cf. [213]. Now we can define a *constraint system*:

Definition 1 (Constraint System). A *constraint system* is a complete algebraic lattice $(\mathcal{C}, \mathcal{D}, \sqsubseteq, \sqcup, true, false)$, where $\mathcal{C}$ is the set of constraints, $\mathcal{D}$ is the set of finite constraints, $\sqsubseteq$ is the partial order, $\sqcup$ is the least upper bound operation, $true$ and $false$ are the bottom and top elements. Two constraints $c, d \in \mathcal{C}$ are said to be *inconsistent* if $c \sqcup d = false$.

In order to model local variables and recursive calls, we need to extend the previous structure with existential quantification. This can be done using the theory of cylindric algebra of Tarski [122]: the constraint system is enriched with a countable set of variables $\mathcal{V}$, cylindrification operators $\exists_x$, $x \in \mathcal{V}$, and diagonal operators d_{xy}, $x, y \in \mathcal{V}$. These new operators are defined through the axioms of a *cylindric constraint system*.

Definition 2. A Cylindric Constraint System $\mathfrak{C} = (\mathcal{C}, \mathcal{D}, \mathcal{V}, \sqsubseteq, \sqcup, true, false, \exists_x, d_{xy})$ is a complete algebraic lattice where $\exists_x$ and d_{xy} satisfy:

- $\exists_x c \sqsubseteq c$;

- if $c \sqsubseteq d$ then $\exists_x c \sqsubseteq \exists_x d$;

- $\exists_x(c \sqcup \exists_x d) = \exists_x c \sqcup \exists_x d$;

- $\exists_x \exists_y c = \exists_y \exists_x c$;

- if $\{c_i\}_{i \in I}$ is an increasing chain, then $\exists_x \bigsqcup_i c_i = \bigsqcup_i \exists_x c_i$;

- $d_{xx} = true$;

- if $z \neq x, y$ then $d_{xy} = \exists_z (d_{xz} \sqcup d_{zy})$;

- id $x \neq y$ then $c \sqsubseteq d_{xy} \sqcup \exists_x (c \sqcup d_{xy})$.

An important instrument while working with constraint systems is the substitutions of variables in a constraint, allowing to model hiding and recursive call in a simple way. In the following, we denote with $fv(c)$ the set of free variables of a constraint c, i.e. all variables appearing in c that are not bounded to any cylindric operator. We can think of a substitution as a mapping $f : \mathcal{V} \rightarrow \mathcal{V}$, with $|\{x \in \mathcal{V} \mid x \neq f(x)\}| < \infty$. Formally, given a constraint c, and two vectors of variables $\mathbf{x} \subset fv(c)$ and $\mathbf{y}$, we define the substitution of $\mathbf{x}$ with $\mathbf{y}$ in c as the constraint $c[\mathbf{x}/\mathbf{y}] = \exists_\alpha (\mathbf{y} = \alpha \sqcup \exists_\mathbf{x}(\alpha = \mathbf{x} \sqcup c))$, with $\alpha \cap (fv(c) \cup \mathbf{y} \cup \mathbf{x}) = \emptyset$ and $|\alpha| = |\mathbf{x}| = |\mathbf{y}|$. Essentially, we need to pass through α variables in order to avoid name clashes. Substitutions defined in this way satisfy the expected properties, cf. [235] for proofs:

- if $c \sqsubseteq d$ then $c[f] \sqsubseteq d[f]$;

- if $f = f'|_{fv(c)}$, then $c[f] = c[f']$;

- $c[f_1][f_2] = c[f_1 \circ f_2]$;

- if f is a renaming, then $\exists_{f(\mathbf{x})} c[f] = (\exists_\mathbf{x} c)[f]$.

There are some subtleties depending on the fact that an unbounded number of α variables is needed to perform the formal linking, and it may be the case that the constraint on which substitution is performed contains all the variables in $\mathcal{V}$. From now on, we will assume that this never happens; the assumption is safe as we can always reserve an infinite number of variables for the substitution itself, cf. [235] for further details.

The operations performed on the constraint store during a computation are of two types: either a constraints c is added or the system is asked to verify if a constraint c is entailed by the current configuration d. In the formalism of constraint systems, the first operation consists in computing the least upper bound of the constraint told and the current status of the store, $c \sqcup d$, while the second operation consists in checking if the relation $d \vdash c$ is true.

Underneath these two operations, there is hidden a mechanism of constraint propagation [10]. The formulas defining the constraint are interpreted on a fixed structure, and each constraint added reduces the permitted domain of the variables of the system. Moreover, in order to compute the relation $\vdash$, we need to know (a representation of) this set of permitted values. However, the presentation of the theoretical foundations of CCP abstract from these details, fundamental for any implementation, moving the focus on the concurrent mechanisms of the language.

$Program ::= Decl.A$

$Decl ::= \varepsilon \mid Decl.Decl \mid p(\mathbf{x}) : -A$

$G ::= \text{tell(c)} \mid \text{ask(c)}$

$M ::= G \rightarrow A \mid M + M$

$A ::= \mathbf{0} \mid A.A \mid M \mid A \parallel A \mid \exists_x A \mid p(\mathbf{x})$

Table 1.1: Syntax of CCP.

1.2   Syntax

The syntax of the language is presented in Table 1.1, as a grammar. A CCP program is composed of two sections: the declaration of procedures and the initial agent to be executed. Procedures are identified by a unique name, and the variables in the declaration of $p(\mathbf{x}) : -A$ must satisfy the condition $\mathbf{x} \supseteq \text{fv(A)}$, where fv(A) denotes the free variables of A, i.e. all the variables not in the scope of an $\exists_x$ operator.

The main actions that an agent can perform are ask and tell. In CCP, they both have the role of a guard in the prefixed choice operator. This is the mechanism used to synchronize processes, as both ask and tell have to satisfy a condition in order to be active; if the condition does not hold, the corresponding agents are suspended. The constraint asked, say c, must be entailed by the current configuration of the store d, i.e. $d \vdash c$ must hold. The told constraint c , instead, must be consistent with the store d, i.e. $c \sqcup d \neq false$. This form of tell is called in literature *atomic tell*, see [213]. There is also another version of tell operator, called *eventual tell* [215], which does not have to pass the consistency check in order to add the constraint to the store. In this case, tell cannot be used as a guard, and therefore it is defined as an agent. Motivations of this difference are to be found in the different denotational semantics that arise from these two kind of tell operators, cf. [68].

Agents prefixed by a guard can be combined together using the non-deterministic choice operator $+$. The resulting non-determinism is a *"don't care"* one, i.e. once a choice has been made, it cannot be undone, and the agent has to *commit* to it. This is in contrast with *"don't know"* non-determinism, typical of non-concurrent logical languages, see [68].

Finally, agents can be either a summation or the null agent and they can be combined sequentially ("." operator) and in parallel ("$\parallel$" operator). We can also declare local variables ("$\exists_x$") and call recursively other procedures.

As a simple example, consider the following CCP program:

$$\exists_X (\text{tell}(X = 1).A_1 \parallel \text{ask}(X = 1).A_2),$$

composed by two agents in parallel synchronizing an a local variable X, and then evolving respectively as A_1 and A_2.

1.3 Operational Semantics

In presenting the operational semantics of CCP, we follow the approach of [69]. As mentioned above, the language is parametric with respect to a constraint system $\mathfrak{C}$, which encodes the peculiarities of the domain of the program being written. At each point of the computation, the store contains all the information produced by the program, in the form of sets of elementary constraints, i.e. subsets of $\mathcal{D}$. As said, computations evolve monotonically, so constraints can only be added to $\mathcal{C}$, but never removed.

The operational semantics of CCP is given in the style of the SOS of Plotkin [192], and it is presented in Table 1.2. The configurations of the system are represented by points in the space $\mathcal{P} \times \mathcal{C}$, where $\mathcal{P}$ is the space of processes and $\mathcal{C}$ is the set of constraints. Therefore, at each point of the computation, a configuration stores the current agent to be executed and the current configuration of the constraint store. Rule (CCP1) deals with the tell operator. It states that when a tell(c) is executed, the constraint c is added to the store d by taking the least upped bound $d \sqcup c$. This operation, however, is conditioned to the satisfaction of the consistency check $c \sqcup d \neq false$. Note that if c is entailed by d, then $d \sqcup c = d$, and the store is not altered (we are not adding new information). Rule (CCP2) presents the semantics of the ask instruction. The computation proceeds only if the asked constraint is entailed by the current configuration of the store. Rule (CCP3) deals with non-deterministic choice: it states that if a choice is enabled, than it can be taken by the system. Notice that this is a blinded committed choice, so there is no possibility of backtracking. Rule (CCP4) deals with the semantics of the parallelism, which is interleaving.
Rule (CCP5) shows how the hiding operator works. The agent $\exists_x^d A$ represents an agent A where the variable x must be considered local, and the information present in d represents the local information generated by A during its past computation (which possibly regards also x). From the internal point of view, this is the usual scoping rule, where the variable x possibly occurring in the global store c is hidden, i.e. the x in c is global, hence "covered" by the local x. Therefore, A has no access to the information on x in c, and this is formally achieved by filtering c with $\exists_x$. Thus, if the store which is visible at the external level is c, then the store which is seen internally by A is $(\exists_x c) \sqcup d$. Now, if A can make a transition, reducing itself to B and transforming the local store into d' , from outside we see that the the agent is transformed into $\exists_x^{d'} B$, and that the information $\exists_x d$ present in the global store is transformed into $\exists_x d'$. In other words, the new information (in particular, the information concerning the local x) is accumulated in the private store of the agent, and the part of it which does not concern the local x is communicated externally.

Finally, rule (CCP6) deals with recursion. The idea is to link the formal parameter **y** to **x** in such a way that all the information present about **x** is passed to **y**. This is done by using the $\Delta_{\mathbf{x}}^{\mathbf{y}} A$ operator, defined for a single variable as $\exists_\alpha^{d_{\alpha y}} \exists_x^{d_{x\alpha}} A$ if $x \neq y$, and as A if $x = y$. What it does is simply renaming y with x in two steps: first y is linked to a fresh variable α, then this new variable is linked to x. Linking is obtained by posting the diagonal constraints $d_{\alpha y}$ and $d_{x\alpha}$. This two-step mechanism is needed to avoid name clashes between variables **x** and **y** when more than one variable needs to be substituted. The operator $\Delta_{\mathbf{x}}^{\mathbf{y}} A$, in fact, is defined as $\Delta_{x_1}^{y_1} \dots \Delta_{x_n}^{y_n} A$.

$$(CCP1) \quad \langle \text{tell}(c) \to A, d \rangle \longrightarrow \langle A, d \sqcup c \rangle \qquad \text{if } c \sqcup d \neq false$$

$$(CCP2) \quad \langle \text{ask}(c) \to A \rangle \longrightarrow \langle A, d \rangle \qquad \text{if } d \vdash c$$

$$(CCP3) \quad \frac{\langle M_1, d \rangle \longrightarrow \langle A, d' \rangle}{\langle M_1 + M_2, d \rangle \longrightarrow \langle A, d' \rangle}$$

$$(CCP4) \quad \frac{\langle A, d \rangle \longrightarrow \langle A', d' \rangle}{\langle A \parallel B, d \rangle \longrightarrow \langle A' \parallel B, d' \rangle}$$

$$(CCP5) \quad \frac{\langle A, d \sqcup \exists_x c \rangle \longrightarrow \langle B, d' \rangle}{\langle \exists_x^d A, c \rangle \longrightarrow \langle \exists_x^{d'} B, c \sqcup \exists_x^{d'} \rangle}$$

$$(CCP6) \quad \langle p(\mathbf{y}), c \rangle \longrightarrow \langle \Delta_{\mathbf{x}}^{\mathbf{y}} A, c \rangle \qquad \text{if } p(\mathbf{x}) : -A \in Decl$$

Table 1.2: Structural Operational Semantics for CCP.

1.4 Probabilistic CCP

In [74], a probabilistic version of CCP (called pCCP) is presented. Probabilities are introduced to solve probabilistically every form of non-determinism, thus allowing to quantify the likelihood of the paths of execution, and consequently reason *quantitatively* about average properties of programs. This is useful, for instance, while dealing with randomized or stochastic algorithms whose behaviour is related to the generation of random numbers, cf. [8, 176]. This probabilistic version of CCP has been studied deeply by their authors, and it has been provided both with a denotational semantics using an algebra of operators [73] and with a mechanism of probabilistic abstract interpretation [75].

The non-determinism in CCP is present in two points, i.e. in the guarded choice and in the interleaving semantics of the parallel operator. The structural semantics of pCCP (see Table 1.3) is modeled with a labeled transition system, where labels represent probabilities associated to the transitions, and probabilities of paths are the product of the probabilities of the single transitions. This transition system can be transformed into a Discrete Time Markov Chain (cf. Chapter 2, Section 2.1), hence the programs of the language have a stochastic behaviour in a discrete time setting. The non-determinism of the choice operator is solved by weighting with a probability distribution the branches of the summation and by choosing a branch with the indicated probability (normalized among active branches). Non-determinism in the interleaving parallel operator is analogously modeled by forcing a probability distribution to the agents in parallel. Both summation and the parallel operator lose associativity with this definition; an example is the following:

$$\frac{1}{2} : A_1 \parallel \frac{1}{2} : (\frac{1}{2} : A_2 \parallel \frac{1}{2} : A_3) = \frac{1}{2} : A_1 \parallel \frac{1}{4} : A_2 \parallel \frac{1}{4} : A_3$$

and

$$\frac{1}{2} : (\frac{1}{2} : A_1 \parallel \frac{1}{2} : A_2) \parallel \frac{1}{2} : A_3 = \frac{1}{4} : A_1 \parallel \frac{1}{4} : A_2 \parallel \frac{1}{2} : A_3.$$

This happens because the probability distribution is attached to the parallel operator, and not to the agents in parallel. In Chapter 6 we present a stochastic version of the language not suffering from this limitation.

A different probabilistic version of CCP is the one presented in [113], where the probabilistic ingredient is introduced adding random variables in the constraint store, without changing the instructions of the language. Therefore, in this approach is the constraint store that becomes stochastic, and not the evolution in terms of interactions. However, this approach is outside the scope of this thesis, so we refer the interested reader to [113] for further details.

1.5 Distributed CCP

A distributed language needs to model execution of agents in a network, composed by several nodes (computational locations) connected by communication links. One important issue of such systems is that communication must be local among nodes. Devising a distributed version of CCP is, therefore, not an easy task, due to the fact that communication between CCP-agents is obtained by posting constraints acting on global variables. Therefore, each communication influences the whole system, and this property inhibits distributed implementations. To tackle this problem, the communication mechanism between CCP agents must be modified, in order to distinguish between modifications of the store acting locally (in a single node of the network) and modifications that involve more than two nodes. Different solutions can be found in literature, all introducing new forms of communication.

In [116], Réty proposes a π-calculus approach to model the communication between agents located in different nodes. In particular, new instructions for exchanging messages synchronously are introduced, even if there is still a single, global, constraint store, and the independence among variables is imposed as a further condition in the definition of the agents. The content of messages sent among (virtual) nodes are abstract constraints whose variables are templates that must be instantiated by the receiving agent. In addition, there is also a linking mechanism that automatically propagates information on chosen variables.

In [23], de Boer et alt. present a similar approach, with global communication performed via synchronous message passing and local communication performed in CCP style. They also introduce a concept of independent local stores, such that the global configuration of a network is the least upper bound of the local constraint stores.

In [156] a different type of synchronous communication is discussed, based on a redefinition of the semantics of ask and tell primitives: a constraint is told only if there in an agent asking for it. Also this version makes use of a single global constraint store.

There is also some work by Palamidessi [97] regarding the migration of CCP agents. CCP is enriched by a hierarchical network structure and agents can move from one node to another one (bringing with them their subagents).

1.6 Stream Variables

In the use of a stochastic variant of CCP as a modeling language for biological systems, many variables will represent quantities that vary over time, like the number of molecules

$$Program ::= Decl.A$$

$$Decl ::= \varepsilon \mid Decl.Decl \mid p(\mathbf{x}) : -A$$

$$G ::= \text{tell(c)} \mid \text{ask(c)}$$

$$A ::= \mathbf{0} \mid A.A \mid p_1 : G_1 \rightarrow A_1 + \ldots + p_n : G_n \rightarrow A_n \mid$$
$$p_1 : A_1 \parallel \ldots \parallel p_n : A_n \mid \exists_x A \mid p(\mathbf{x})$$

$(pCCP1)$
$$\langle \text{tell}(c) \rightarrow A, d \rangle \longrightarrow \langle A, d \sqcup c \rangle$$
$$\text{if } c \sqcup d \neq false$$

$(pCCP2)$
$$\langle \text{ask(c)} \rightarrow A \rangle \longrightarrow \langle A, d \rangle$$
$$\text{if } d \vdash c$$

$(pCCP3)$
$$\frac{\langle G_1 \rightarrow A_1, d \rangle \longrightarrow \langle A_1, d' \rangle}{\langle p_1 : G_1 \rightarrow A_1 + \ldots + p_n : G_n \rightarrow A_n, d \rangle \longrightarrow_{\widetilde{p_1}} \langle A, d' \rangle}$$
$$\text{where } \widetilde{p_1} = p_1 / \sum_{i:G_i \ active} p_i$$

$(pCCP4)$
$$\frac{\langle A_1, d \rangle \longrightarrow_p \langle A_1', d' \rangle}{\langle p_1 : A_1 \parallel \ldots \parallel p_n : A_n, d \rangle \longrightarrow_{p \cdot \widetilde{p_1}} \langle p_1 : A_1' \parallel \ldots \parallel p_n : A_n, d' \rangle}$$
$$\text{where } \widetilde{p_1} = p_1 / \sum_{i:A_i \ active} p_i$$

$(pCCP5)$
$$\frac{\langle A, d \sqcup \exists_x c \rangle \longrightarrow_p \langle B, d' \rangle}{\langle \exists_x^d A, c \rangle \longrightarrow_p \langle \exists_x^{d'} B, c \sqcup \exists_x^{d'} \rangle}$$

$(pCCP6)$
$$\langle p(\mathbf{y}), c \rangle \longrightarrow_1 \langle \Delta_{\mathbf{x}}^{\mathbf{y}} A, c \rangle$$
$$\text{if } p(\mathbf{x}) : -A \in Decl$$

Table 1.3: Syntax and Structural Operational Semantics for probabilistic CCP. The language is extended by simply adding a probability distribution to each choice and each parallel composition. The number of summands and of parallel agents must be expressed explicitly, due to lack of associativity of these operators. We require that, for each operator, $\sum_i p_i = 1$, hence the p_is form a probability distribution. The SOS is given by a labeled transition relation $\longrightarrow \subseteq (\mathcal{P} \times \mathcal{C}) \times [0,1] \times (\mathcal{P} \times \mathcal{C})$, using the transition relation of CCP for guards, whose semantics remains unaltered. The most interesting rules are the ones dealing with choice and parallel composition: the system chooses one of the active instruction after renormalizing the probability distribution among active agents. Notice that the renormalization performed at this level makes the condition $\sum_i p_i = 1$ redundant. The last two rules are similar to the ones for CCP, as they are not involved in probabilistic branching.

of certain chemical species. Unfortunately, the variables we have at our disposal in CCP are rigid, in the sense that, whenever they are instantiated, they keep that value forever. However, time-varying variables can be easily modeled as growing lists with an unbounded tail: $X = [a_1, \ldots, a_n | T]$. When the quantity changes, we simply need to add the new value, say b, at the end of the list by replacing the old tail variable with a list containing b and a new tail variable: $T = [b | T']$. When we need to compute a function depending on the current value of the variable X, we need to extract from the list the value immediately preceding the unbounded tail. This can be done by defining the appropriate predicates in the first-order language on which the constraint store is built. As these variables have a special status in the presentation hereafter, we refer to them as *stream variables*. In addition, we will use a simplified notation that hides all the details related to the list update. For instance, if we want to add 1 to the current value of the stream variable X, we will simply write $X = X + 1$. The intended meaning of this notation is clearly: "extract the last ground element n in the list X, consider its successor $n + 1$ and add it to the list (instantiating the old tail variable as a list containing the new ground element and a new tail variable)". In the rest of the thesis, we will always make clear when a variable must be considered as a stream variable or as a normal one.

Chapter 2

Stochastic Processes

In this chapter we recall briefly definitions and properties of a family of stochastic processes known as Markovian. Their main characteristic is that they retain no memory of what happened in the past, hence their behaviour is determined just by their current state. This *memoryless* feature makes them attractive as models of stochastic computations. In this thesis, we will deal only with state space with a finite or countable number of elements. Whenever this is the case, a Markov process takes the name of *Markov Chain*. We will consider chains both in *discrete time* $\mathbb{N}$ (Discrete Time Markov Chains, DTMC for short) and in *continuous time* $\mathbb{R}^+$ (Continuous Time Markov Chains, CTMC for short). Our introduction will be concise; for a tough and extended presentation about Markov Chains we refer to the book of Norris [181]. In addition, we suppose the reader to be familiar with concepts of probability theory, referring him to an introductory book like [18].

In Section 2.1 we present DTMC. After setting the basic concepts (Section 2.1.1), we focus our attention on class structure, absorption probabilities (Section 2.1.2), and ergodic properties (Section 2.1.3). After dealing with discrete time processes, we shift the attention to continuous-time processes first discussing exponential distributions (Section 2.2). Then, we present the basics of CTMC, with particular emphasis on Q-matrices (Section 2.3), jumping chains and holding times (Section 2.3.1). Successively, we will give a brief overview of class structure, absorption probabilities (Section 2.3.2), and ergodic properties (Section 2.3.3) also in the continuous time case. Finally, in Section 2.4 we briefly present some methods to simulate the probability distributions we are interested in, while in Section 2.5 we present some basic concepts about model checking of Continuous Time Markov Chains.

2.1 Discrete Time Markov Chains

2.1.1 Basics

Let S denote the state space, supposed to be finite or countable[1]. We denote with λ a probability distribution on S, thus $\lambda = (\lambda_i)_{s_i \in S}$ and $\sum_{s_i \in S} \lambda_i = 1$. With Π, instead, we denote a stochastic matrix on S, i.e. a matrix $\Pi = (\pi_{ij})_{s_i, s_j \in S}$ such that each row sums up to one, $\sum_{s_j \in S} \pi_{ij} = 1$. Finally, with $X : \Omega \to S$ we denote a generic random variable on S.

---

[1]We denote elements of S as s_i, s_j, indexing the corresponding elements on vectors on S by i, j.

Definition 3. A sequence $(X_n)_{n \in \mathbb{N}}$ of random variables on S is called a (Discrete Time) Markov Chain with initial distribution λ and transition matrix Π if

1. $\mathbb{P}(X_0 = s_0) = \lambda_0$

2. $\mathbb{P}(X_{n+1} = s_{i_{n+1}} | X_0 = s_{i_0}, \ldots, X_n = s_{i_n}) = \pi_{i_n i_{n+1}}$.

For short, we indicate the property of being a Markov Chain as Markov(λ, Π). The condition *(2)* in the previous definition is the memoryless property of Markov Chains: the probability of going to state $s_{i_{n+1}}$ depends only on the current state s_{i_n}. The probability distribution (p.d. for short) λ identifies the initial state of the chain. Sometimes it is more interesting to consider the evolution of the chain starting from a fixed state, i.e. the case in which λ is unimodal. We denote such p.d. with $\delta_i = (\delta_{ij})_{s_i s_j \in S}$, where δ is the usual δ of Kronecker. Moreover, with $\mathbb{P}_i(A)$ we denote the probability conditional to $X_0 = s_i$, i.e. $\mathbb{P}(A | X_0 = s_i)$.

Two basic properties of a Markov Chain are given by the following

Theorem 1. Let $(X_n)_{n \in \mathbb{N}}$ be Markov(λ, Π). Then

1. $\mathbb{P}(X_0 = s_{i_0}, \ldots, X_n = s_{i_n}) = \lambda_{i_0} \pi_{i_0 i_1} \cdots \pi_{i_{n-1} i_n}$

2. Conditional on $X_m = s_i$, $(X_{m+n})_{n \in \mathbb{N}}$ is Markov(δ_i, Π) and it is independent on $X_1, \ldots, X_m$.

The property *(1)* of the theorem gives a way to compute probability of traces of the chain, while property *(2)*, known usually as Markov Property, states clearly the absence of memory of Markov Chains. Proof of this theorem can be found in [181].

When the state space S is finite, so is the dimension of vectors λ and stochastic matrices Π. In this case, the product of matrices can be defined in the standard way, and thus with Π^n we denote the n-th power of matrix Π. A similar definition can be given also for infinite stochastic matrices, thanks to their property of having a bounded finite sum for each row. We skip from these details here, referring the reader to [218]. The following theorem establish a link between the probability of being in a state after n steps and the multiplication of matrices.

Theorem 2. Let $(X_n)_{n \in \mathbb{N}}$ be Markov(λ, Π). Then, for each $n, m \geq 0$

1. $\mathbb{P}(X_n = s_j) = (\lambda \Pi^n)_j$

2. $\mathbb{P}_i(X_n = s_j) = \mathbb{P}_i(X_{n+m} = s_j | X_m = s_i) = \pi_{ij}^n$.

2.1.2 Class Structure and Absorption Probabilities

A fruitful way of representing a DTMC, particularly in the case of finite state spaces S, is as a weighted graph. More precisely, the set of vertices of the graph coincides with the state space S, while the weighting function for the edges is given by $w : S \times S \to \mathbb{R}$, $w(s_i, s_j) = \pi_{ij}$. The *support graph* is a directed graph with vertices in S and edges $\{(s_i, s_j) \mid \pi_{ij} > 0\}$. It is clear from the definition that we can reach a state s_j starting from s_i with non-zero probability if and only if there exists a path from s_i to s_j in the support graph of the Markov Chain. We denote this property with $s_i \longrightarrow s_j$. The following theorem is straightforward:

Theorem 3. *Given two states $s_i \neq s_j$, the following conditions are equivalent:*

1. $s_i \to s_j$.

2. *There exists a sequence of states $s_{i_0}, s_{i_1}, \ldots, s_{i_n}$ with $s_{i_0} = s_i$, $s_{i_n} = s_j$ and $\pi_{i_0 i_1} \cdots \pi_{i_{n-1} i_n} > 0$.*

3. $\pi_{ij}^n > 0$ *for some n.*

If $s_i \to s_j$ and $s_j \to s_i$, we write $s_i \leftrightarrow s_j$. $\leftrightarrow$ is an equivalence relation, and partition the state space S into *communicating classes*. A communicating class C is *closed* whenever $s_i \in C$ and $s_i \to s_j$ implies that $s_j \in C$. If a class contains just one element, it is called *absorbing*. Thinking at the characterization of $\to$ in terms of the support graph, we can easily realize that communicating classes correspond to *strongly connected components* (s.c.c.) of the support graph. Considering the graph of the strongly connected components, we can easily see that closed classes correspond to s.c.c. with no exiting transitions.

A state s_i is called *transient* if $\mathbb{P}_i(X_n = s_i$ for infinitely many $n) = 0$, while it is called *recurrent* if the previous probability is 1. For Markov Chains, the *dichotomy* property holds: a state is either recurrent or transient. If the state space S is finite, the recurrent states are exactly those belonging to closed classes. Taking the point of view of s.c.c. graph, it is reasonable that a s.c.c. that is not absorbing will be eventually abandoned forever, and that at a certain point the process will be "trapped" in a closed class.

Once the class structure of a chain is known, we may ask what is the probability of hitting a certain state or set of states, and what is the average time we have to wait for this to happen. The *hitting time* for $A \subseteq S$ is a random variable $H^A : \Omega \to \{0, 1, 2, \ldots\} \cup \{\infty\}$ defined by

$$H^A(\omega) = \inf_{n \geq 0}\{X_n(\omega) \in A\}.$$

The *hitting probability* of A is the probability of ever hitting A starting from a state s_i:

$$h_i^A = \mathbb{P}_i(H^A < \infty).$$

If A is a closed class, then h_i^A is also called *absorption probability*. Finally, the *average hitting time* of A is $k_i^A = \mathbb{E}_i(H^A)$.

Hitting probability and average hitting times can be computed explicitly by solving systems of linear equations, as stated in the following

Theorem 4. *The vector of hitting probabilities $(h_i^A)_{s_i \in S}$ is the minimal non-negative solution of the system of linear equations:*

$$\begin{cases} h_i^A = 1 & \text{for } s_i \in A \\ h_i^A = \sum_{s_j \in S} p_{ij} h_j^A & \text{for } s_i \notin A \end{cases}$$

The vector of average hitting times $(k_i^A)_{s_i \in S}$ is the minimal non-negative solution of the system of linear equations:

$$\begin{cases} k_i^A = 0 & \text{for } s_i \in A \\ k_i^A = \sum_{s_j \notin A} p_{ij} k_j^A & \text{for } s_i \notin A \end{cases}$$

2.1.3   Invariant Distributions and Ergodicity

The long term behaviour of a Markov Chain is related to the existence of *invariant measures*, i.e. probability vectors λ such that $\lambda\Pi = \lambda$. Such distributions are also called *stationary* or *equilibrium*. These names are justified by the following

Theorem 5. *Let $(X_n)_{n\in\mathbb{N}}$ be Markov(λ, Π). Then*

    1. *If λ is invariant for Π, then $(X_{m+n})_{n\in\mathbb{N}}$ is also Markov(λ, Π) (stationarity).*

    2. *If S is finite and for some $s_i \in S$ and all $s_j \in S$, $\pi_{ij}^n \to \mu_j$ for $n \to \infty$, then $\mu = (\mu_j)_{s_j\in S}$ is an invariant distribution (equilibrium).*

The existence of an invariant measure is guaranteed if a Markov Chain is *ergodic* or *irreducible*. This property states that we can reach with non-zero probability any state j starting from any state i, after a finite number of steps. Equivalently, a chain is ergodic if the support graph is strongly connected. A state of the chain is *positive recurrent* whenever its expected return time is finite. The *return time* for state s_i is a random variable $T_i : \Omega \to \mathbb{N}\cup\{\infty\}$ defined by $T_i = \min\{n > 0 \mid X_n = s_i$ given that $X_0 = s_i\}$, and the *expected return time* is $m_i = \mathbb{E}_i(T_i)$. The following theorem gives conditions sufficient for the existence of an invariant measure.

Theorem 6. *Let Π be irreducible. the following statements are equivalent:*

    1. *Every state in S is positive recurrent.*

    2. *Some state $s_i \in S$ is positive recurrent.*

    3. *π has an invariant distribution μ. In this case, $m_i = 1/\mu_i$.*

If we impose a stronger condition than positive recurrence, namely aperiodicity, we can prove a result on the limit behaviour. A state $s_i \in S$ is *aperiodic* if $\pi_{ii}^n > 0$ ultimately (i.e. for all $n \geq n_0$). A chain is aperiodic if so are all its states.

Theorem 7 (Convergence to Equilibrium). *Let Π be irreducible and aperiodic, and let μ be an invariant distribution for Π. Let $(X_n)_{n\in\mathbb{N}}$ be Markov(λ, Π) for an arbitrary distribution λ. Then*

$$\forall s_j \in S, \ \mathbb{P}(X_n = s_j) \to \mu_j, \ n \to \infty.$$

In particular,

$$\forall s_i, s_j \in S, \ \pi_{ij}^n \to \mu_j, \ n \to \infty.$$

2.2   Exponential Distributions

Consider a random variable $T : \Omega \longrightarrow [0,\infty]$. We say that T has an *exponential distribution of parameter (or rate) λ, $\lambda \in [0,\infty)$ ($T \sim Exp(\lambda)$ for short)*, if its probability function is

$$\mathbb{P}(T > t) = e^{-\lambda t} \text{ for all } t \geq 0.$$

The density function of T is

$$f_T(t) = \lambda e^{-\lambda t}\chi_{\{t\geq 0\}}.$$

where χ_A denoted the characteristic function of set A. The expected value of T is

$$\mathbb{E}(T) = \int_0^\infty \mathbb{P}(T > t)dt = \frac{1}{\lambda}.$$

The crucial property of exponential distributions in the modeling of computations is stated in the following

Theorem 8 (Memoryless Property). *A random variable $T : \Omega \longrightarrow (0, \infty]$ has an exponential distribution if and only if the following memoryless property holds:*

$$\mathbb{P}(T > s + t | T > s) = \mathbb{P}(T > t) \text{ for all } s, t \geq 0.$$

Proof is omitted and can be found in [181].

Suppose we have some random processes governed by an exponential distribution, competing in a race condition in such a way that the fastest is executed. We can model this situation with two random variables: one (T) returning the infimum of the random times of all processes, and the other (K) returning the process realizing this infimum. Their stochastic behaviour is described in the following

Theorem 9. *Let I be a countable set and let T_k, $k \in I$, be independent random variables, with $T_k \sim Exp(q_k)$ and $q = \sum_{k \in I} q_k < \infty$. Set $T = \inf_k T_k$. Then this infimum is obtained at a unique random value K of k, with probability 1. Moreover, T and K are independent, $T \sim Exp(q)$ and $\mathbb{P}(K = k) = q_k/q$.*

Proof. Set $K = k$ if $T_k < T_j$ for all $j \neq k$, K is undefined otherwise. Then

$$\begin{aligned}
\mathbb{P}(K = k \text{ and } T \geq t) &= \mathbb{P}(T_k \geq t \text{ and } T_j > T_k \text{ for all} j \neq k)\\
&= \int_t^\infty q_k e^{-q_k s} \mathbb{P}(T_j > s \text{ for all} j \neq k)ds\\
&\int_t^\infty q_k e^{-q_k s} \prod_{j \neq k} e^{-q_j s}ds\\
&\int_t^\infty q_k e^{-qs}ds = \frac{q_k}{q}e^{-qt}
\end{aligned}$$

Computing the marginal distributions for K and T, we obtain the claimed results. Moreover, their joint distribution turns out to be the product of the marginals, thus showing that K and T are independent and that $\mathbb{P}(K = k \text{ for some } k) = 1$.

2.3 Continuous Time Markov Chains

Let S be the set of states, finite or countable as in the discrete case. A *continuous-time random process*

$$(X_t)_{t \geq 0} = \{X_t \mid t \geq 0\},$$

with values in S, is a family of random variables $X_t : \Omega \longrightarrow S$ that are *right-continuous* w.r.t. t. This means that for all $\omega \in \Omega$ and $t \geq 0$, there exists $\varepsilon > 0$ such that $X_s(\omega) = X_t(\omega)$, for all $t \leq s \leq t + \varepsilon$. Right continuous processes are determined by their finite-dimensional distributions $\mathbb{P}(X_{t_0} = s_0, \ldots, X_{t_n} = s_n)$, for $0 \leq t_0 \leq \ldots t_n$ and $s_0, \ldots, s_n \in S$.[2]

---

[2]All the probabilities of interest can be computed by taking the limit for $n \longrightarrow \infty$ of finite-dimensional distributions, letting t_i take values on an enumeration of rationals.

A *Continuous Time Markov Chain* is a particular right-continuous continuous-time random process, determined by the *memoryless condition*, holding true for all different n, t_i and s_i:

$$\mathbb{P}(X_{t_n} = s_n \mid X_{t_0} = s_0, \ldots, X_{t_{n-1}} = s_{n-1}) = \mathbb{P}(X_{t_n} = s_n \mid X_{t_{n-1}} = s_{n-1}).$$

CTMC can be described using through their *generator matrix*, or *Q-matrix*, which is a matrix whose rows and columns are indexed by S, and whose entry $q_{ij} \geq 0$, $i \neq j$ represents the rate of an exponential distribution associated to the transition from state s_i to state s_j. The value q_{ii} is set equal $-\sum_{j \neq i} q_{ij}$ (assumed to be finite for each i), so that each row of the Q-matrix sums up to zero. The definitions via memoryless condition and via Q-matrices can be proved equivalent, cf. [181]. A good way to visualize a CTMC is to associate to it a complete graph, whose nodes are indexed with elements of S, and whose edges are labeled with the corresponding rate of the Q-matrix (edges with rate zero can be omitted).

Q-matrices determine the stochastic behaviour of the chain. In fact, they induce a semi-group of sub-stochastic matrices[3] $(P(t) : t \geq 0)$, such that $P(t_1)P(t_2) = P(t_1 + t_2)$ for all $t_1, t_2 \geq 0$. The value $p_{ij}(t)$ represents the probability of going from state s_i to state s_j in t units of time. Matrices $P(t)$ can be computed as the unique solution of the following *forward equation*:

$$\frac{d}{dt}P(t) = P(t)Q, \quad P(0) = Id,$$

or equivalently as the unique solution of the *backward equation*:

$$\frac{d}{dt}P(t) = QP(t), \quad P(0) = Id.$$

In the case of S finite, we can give a representation of $P(t)$ in terms of *matrix exponentiation*:

$$P(t) = e^{tQ},$$

where e^{tQ} is define as the componentwise limit of the series

$$\sum_{n=0}^{\infty} \frac{Q^n}{n!}.$$

Further details can be found again in [181].

The memoryless condition of CTMC can be described equivalently as a race condition. We consider all the positive entries in the generator matrix Q, associating to each $q_{ij} > 0$ an exponentially distributed random variable $T_{ij} \sim Exp(q_{ij})$. When the chain is in state s_i, then a race condition begins among all random variables T_{ij}, $s_j \in S$, $q_{ij} > 0$. Notice that each random variable T_{ij} correspond to an edge in the support graph associated to the CTMC. The race is won by the fastest variable, i.e. the one realizing $T_i = \inf_{s_j \in S}\{T_{ij}\}$. If T_{ik} is such variable, then the system moves in state s_k in $T = T_{ik}$ units of time. When the system reaches state s_k, then a race condition between random variables exiting from s_k begins anew.

---

[3]In a sub-stochastic matrix all rows are semi-positive and their sum is less or equal to one.

2.3.1 Jump Chains and Holding Times

CTMC can be described also in terms of their jump chains and holding times. Intuitively, the behavior of a CTMC is a sequence of instantaneous "state jumps", interleaved by a "rest time" spent in each state. To formalize the mechanism above, we can proceed in the following way. Let $(X_t)_{t\geq 0}$ be a CTMC; the *jump times* are a sequence $J_0, J_1, \ldots$ of random variables with values in $(0, \infty)$ defined inductively by

$$J_0 = 0; \quad J_{n+1} = \inf\{t \geq J_n \mid X_t \neq X_{J_n}\}.^4$$

The *holding times* represent the time spent in each state, and are defined in terms of $(J_i)_{i\geq 0}$ as

$$S_n = \begin{cases} J_{n+1} - J_n & \text{if } J_n < \infty \\ \infty & \text{otherwise} \end{cases}$$

The jump times represent the temporal instants when a jump from a state to another occurs. We can discard the temporal information and remember only the sequence of visited stated, thus defining a discrete-time Markov chain $(Y_n)_{n\geq 0}$, $Y_n = X_{J_n}$, the so-called *jump chain*.

The memoryless condition of CTMC, expressed in terms of race conditions and combined with Theorem 9, attaches to jump chains and holding times a simple stochastic behaviour, that is described in the following

Theorem 10. *Let $(X_t)_{t\geq 0}$ be a CTMC, and let Q be its Q-matrix. Call $q_i = -q_{ii} = \sum_{j\neq i} q_{ij}$. Then*

1. *The jump chain $(Y_n)_{n\geq 0}$ is a discrete-time Markov chain with stochastic matrix Π defined in terms of the Q-matrix Q as $\pi_{ij} = \frac{q_{ij}}{q_i}$ for $i \neq j$ and $\pi_{ii} = 0$;*

2. *Conditional on $Y_0 = s_{i_0}, \ldots, Y_n = s_{i_n}$, the holding times $S_0, \ldots, S_n$ are independent exponential random distributions of rates $q_{i_0}, \ldots, q_{i_n}$.*

Therefore, in a CTMC the discrete sequence of state jumps is governed by the stochastic matrix of normalized rates, while the time spent in a single state is determined by the sum of the rates of the transitions exiting from it (the *exit rate*).

2.3.2 Class Structure and Absorption Probabilities

Many properties of CTMC are connected with the discrete-time jump process $(Y_n)_{n\geq 0}$. One of those properties is the class structure of the process. This is not surprising, as this property depends on the support graph, and the support graph for $(X_t)_{t\geq 0}$ and $(Y_n)_{n\geq 0}$ are the same. The notion of communicating states, classes, and absorbing classes are defined like for DTMC. The classification of states in recurrent and transient is also determined by the jump chain.

If class structure and recurrence are determined by the jump chain, the hitting time and hitting probability of a CTMC are determined by the jump times. Given a CTMC $(X_t)_{t\geq 0}$ with generator matrix Q, the hitting time of $A \subset S$ is a random variable D^A defined by

$$D^A(\omega) = \inf\{t \geq 0 \mid X_t(\omega) \in A\}.$$

---

[4]We agree that $\inf \emptyset = \infty$.

If H^A is the hitting time for the jump chain $(Y_n)_{n \geq 0}$, see Section 2.1.2, then it holds that $D^A = J_{H^A}$, an equation connecting D^A with the jump times. From this property, it follows that the hitting probabilities for the CTMC, i.e. $h_i^A = \mathbb{P}_i(D^A < \infty)$, coincide with that of the jump chain and can be calculated in the same way.

2.3.3 Invariant Distributions and Ergodicity

Ergodicity, or irreducibility, for CTMC is defined analogously to DTMC (Section 2.1.3): the support graph has to be strongly connected. A distribution λ is said to be *invariant* for a Q-matrix Q whenever $\lambda Q = 0$. If Π is the matrix of the jump chain, and μ is defined by $\mu_i = \lambda_i q_i$, then invariance can be proven equivalent to μ being an eigenvector of Π with eigenvalue 1, i.e. $\mu \Pi = \mu$. Also for CTMC, irreducibility and recurrence can be shown to be sufficient conditions for the existence of an invariant distribution, unique up to scalar multiples.

The existence of an invariant measure for ergodic Q-matrices determines, as in the discrete case, the limit behaviour of the chain:

Theorem 11 (Convergence to Equilibrium). *Let Q be an irreducible[5] Q-matrix with semigroup $P(t)$, having an invariant distribution λ. For all states $s_i, s_j \in S$, it holds that*

$$p_{ij}(t) \to \lambda_j, \ t \to \infty.$$

2.4 Stochastic Simulation

In this section we present quickly some basic methods used in stochastic simulation. The problem faced is that of sampling from a know probability distribution, either continuous or discrete. The starting point of all such methods is some algorithmic procedure capable of generating pseudo-random numbers. Usually, such generators produce a real random number uniformly distributed in $[0, 1]$, see [206]. In the following we denote by $U(a, b)$ a uniform random variable with support in $[a, b]$, and with u a single realization of such variable. Our random number generators return numbers $u \in [0, 1]$, drawn according to $U(0, 1)$.

Simulating a discrete distribution. Suppose X is a discrete random variable on a finite space $S = \{1, \ldots, n\}$, with probability distribution $\{p_1, \ldots, p_n\}$. To sample from S, we simply need to draw a random number u, and then find the smallest index k such that $\sum_{i=1}^{k} p_i \geq u$. To see why this method works, call $q_k = \sum_{i=1}^{k} p_i = \mathbb{P}(X \leq k)$; as to sample k we must have $q_{k-1} \leq U \leq q_k$, then

$$\mathbb{P}(X = k) = \mathbb{P}(U \in (q_{k-1}, q_k]) = q_k - q_{k-1} = p_k.$$

Inverse distribution method. The following theorem gives a straightforward method to simulate a class of simple continuous probability distributions.

---

[5]The matrix must also be non-explosive, meaning that there cannot happen a infinite number of transitions in a finite time.

Theorem 12. *If U is a random variable with probability distribution $U(0,1)$ (i.e. uniformly distributed in $(0,1)$), and F is a valid invertible cumulative distribution function, then*

$$X = F^{-1}(U)$$

has cumulative distribution function F.

Proof. We denote with F_U the cumulative distribution function of U, i.e. $F_U(u) = \mathbb{P}(U \leq u) = u$, $0 \leq u \leq 1$. Then

$$
\begin{aligned}
\mathbb{P}(X \leq x) &= \mathbb{P}(F^{-1}(U) \leq x) \\
&= \mathbb{P}(U \leq F(x)) \\
&= F_U(F(x)) \\
&= F(x).
\end{aligned}
$$

∎

Simulating an exponential distribution. We can apply Theorem 12 to simulate an exponentially distributed random variable. If U is a uniform random variable $U(0,1)$ and $\lambda > 0$, then $X = -\frac{1}{\lambda}\log(U)$ is $Exp(\lambda)$. In fact, $Exp(\lambda)$ has cumulative probability distribution given by $F(x) = 1 - e^{\lambda x}$, invertible with inverse $F^{-1}(u) = -\frac{1}{\lambda}\log(u-1)$.

2.5 Model Checking Continuous Time Markov Chains

In this section we provide a brief overview of model checking methodologies for Continuous Time Markov Processes. In order for model checking algorithms to be decidable, we must restrict to finite state spaces, and to Markov Chains with Q-matrices taking values in $\mathbb{Q}$. We first give an idea of how a measure of probability along paths, or traces, can be defined. Then we present labeled CTMC and the logic used for queries, namely Continuous Stochastic Logic, CSL. In the final section, we give some hints on the algorithms used for model checking and their complexity.

2.5.1 Traces and Probability Measures

Let $(X_t)_{t \geq 0}$ be a CTMC on S, and let Q be its Q-matrix. A trace τ is a non-empty sequence $s_{i_0} t_0 s_{i_1} t_1 s_{i_2} t_2 s_{i_3} \ldots$, where $q_{i_j i_{j+1}} > 0$ for all $j \in \mathbb{N}$ and t_j denotes the time spent in state s_{i_j}. The set of traces is indicated by $\mathcal{T}$. We can define a probability measure on the set $\mathcal{T}$ in the following way. We fix $n+1$ states $s_{i_0}, \ldots, s_{i_n}$, with $q_{i_j i_{j+1}} > 0$, and n non-empty and non-negative real intervals $I_0, \ldots, I_{n-1}$. Cylindric sets are defined as $C(s_{i_0}, I_0, \ldots, I_{n-1}, s_{i_n}) = \{s'_{i_0} t_0 s'_{i_1} t_1 s'_{i_2} \ldots \mid \text{where } s'_{i_j} = s_{i_j} \text{ for } j = 0, \ldots, n \text{ and } t_j \in I_j, \text{ for } j = 0, \ldots, n-1\}$; they form a σ-algebra. The probability of such a cylindric set is defined recursively as

$$
\begin{aligned}
\mathbb{P}(C(s_{i_0}, I_0, \ldots, I_{n-1}, s_{i_n})) &= \mathbb{P}(C(s_{i_0}, I_0, \ldots, I_{n-2}, s_{i_{n-1}}))\mathbb{P}(s_{i_{n-1}}, I_{n-1}, s_{i_n}); \\
\mathbb{P}(s_{i_{n-1}}, I_{n-1}, s_{i_n}) &= \pi_{i_{n-1} i_n}\left(e^{-\inf(I_n)\cdot(-q_{i_n i_n})} - e^{-\sup(I_n)\cdot(-q_{i_n i_n})}\right),
\end{aligned}
$$

where $\inf(I_n)$ and $\sup(I_n)$ denote respectively the left and right bound of interval I_n.

2.5.2 Labeled Continuous Time Markov Chains

In order to query CTMC, we need to add some information. This is done, in the usual way, by labeling each state of the chain with a set of atomic propositions, taken from AP, true in that state. This is done by a labeling function $L : S \rightarrow 2^{AP}$. In addition, we define also two cost functions, allowing us to talk about expected cost of satisfying a formula. The first function $C : S \times S \rightarrow \mathbb{R}^+$ is called the *instantaneous cost function*, returning the cost associated to a transition, while the second function, $c : S \rightarrow \mathbb{R}^+$, is the *cumulative cost function*, corresponding to the cost per unit of time of staying in a state. Hence, if we spend t units of time in state $s \in$ S, the corresponding cost will be $c(s) \cdot t$.

Given the functions C and c, we can define the cost of reaching a set belonging to $F \subset S$ along trace τ, $cost(F)(\tau)$, as ∞ is τ does not contain states of F, or as the sum of the instantaneous and cumulative costs of τ up to the first state of F in τ.

2.5.3 Continuous Stochastic Logic

Continuous Stochastic Logic is used to specify properties of CTMC. It has been introduced in [11] and extended in [14]. Its syntax is defined by the following grammar:

$$
\begin{aligned}
\varphi &::= true|a|\neg\varphi|\varphi \wedge \varphi|\mathcal{P}_{op}[\psi]|\mathcal{S}_{op}[\varphi]|\mathcal{E}_{oc}[\varphi] \\
\psi &::= \mathbf{X}\varphi|\varphi\mathbf{U}^I\varphi|\varphi\mathbf{U}\varphi,
\end{aligned}
\tag{2.1}
$$

where $a \in AP$ is an atomic proposition, $\circ \in \{\leq, <, >, \geq\}$, $p \in [0, 1]$, $c \in \mathbb{R}^+$, and I is an interval of $\mathbb{R}^+$.

Formulae of CSL are divided into state formulae φ and path formulae ψ. State formulae can be atomic propositions, conjunctions and negations of other state formulas, or probabilistic statements. $\mathcal{P}_{op}[\psi]$ indicates that the path formula ψ being satisfied meets the given bound $\circ p$. The probability of a path formula ψ is the probability of the set of traces that satisfy ψ, which can be shown to belong to the cylindric σ-algebra defined above. $\mathcal{S}_{op}[\varphi]$, instead, indicates that the steady state probability of a state formula φ, namely the stationary probability of the set of states satisfying φ, meets the given bound $\circ p$. Finally, $\mathcal{E}_{oc}[\varphi]$ states that the expected cost to reach a state satisfying ψ from the initial configuration meets the bound $\circ c$.

Path formulas, instead, are defined using temporal operators *next* ($\mathbf{X}$), *until* ($\mathbf{U}$), and *time-bounded until* ($\mathbf{U}^I$). The meaning of $\mathbf{X}\varphi$ is that in one transition we reach a state satisfying ψ. $\varphi_1\mathbf{U}\varphi_2$, instead, states that along the current path, the formula φ_1 is true until the formula φ_2 becomes true. In this version of the until operator, no limit is imposed to the time when the path reaches a state satisfying ψ_2, while in the time-bounded until $\mathbf{U}^I$ this event must happen in a time $t \in I$. As routine in temporal logic, we can define the box operator $\square$ (globally) and the diamond operator $\Diamond$ (eventually) from next and until, see, for instance, [55].

The semantics of CSL is simply defined in a recursive manner; we refer the interested reader to [211] for a quick introduction, or to [11, 14] for a deeper discussion.

2.5.4 Model Checking Algorithms

Model checking algorithms for CSL test if a given CTMC satisfy a given state formula φ. Different kind of computations are requested to deal with the three operators $\mathcal{P}$, $\mathcal{S}$, and $\mathcal{E}$. Checking the satisfiability of a $\mathcal{P}(\psi)$ formula requires to compute the probability of the set of paths satisfying ψ. Next and unbounded until formulae have a probability independent on the elapsed time, hence the corresponding computation depends only on the underlying discrete time chain. For the until operator, probability can be computed solving a linear system of the size of the state space. Time-bounded until, instead, requires more sophisticated methods, as we need to compute the transient probability $P(t)$ of the chain. The satisfiability of a $\mathcal{S}(\varphi)$ formula can be dealt by computing the steady state probability of the chain, and then summing up the steady state probability of states satisfying φ. Finally, formulae $\mathcal{E}(\varphi)$ can be checked by solving a linear system of size equal to the state space. Hence, model checking of CSL formulae has a polynomial complexity w.r.t. the size of the search space, depending essentially on the complexity of the method employed to solve the linear systems. We refer the interested reader to [211] for further details.

2.5.5 PRISM

Probably, the most widespread program supporting model checking of CSL formulae over continuous time Markov models is PRISM [155], implementing also model checking procedures for discrete time chains and other variants.

In this thesis, we will use this model checker in order to analyze some (simple) biological systems. In the following, we present briefly the language used by PRISM to describe the CTMC models that are checked; further details can be found in the PRISM user manual, downloadable together with the program [1].

The two basic concepts of the PRISM language are modules and variables. Modules are simply processes interacting together through global shared variables. Each module can also declare some local variables, that determine its inner state. Variables can take integer values, they must have a bounded domain, and they must be initialized with suitable initial values. The instructions of the language are all of the following kind:

```
[]  guard  ->  rate_1:update_1 + ... + rate_n:update_n
```

Guards are conjunctions of atomic propositions built using equality and inequality predicates, and the usual arithmetic functions. If a guard is satisfied, then there is a race condition between n different instructions, each with its own rate `rate_i`. Such instructions can update the value of the variables accessible from the module, and are conjunctions of statements of the form `x' = expr`, where the prime indicates the new value of variable `x`. There is also another source of non-determinism, namely the fact that different guards can be satisfied by a given configuration of the system, hence each module has a competition among all different update instructions corresponding to active guards.

Chapter 3

Mathematical Modeling of Biological Systems

Molecular biology, in all its *"omics"* variants, like *genomics* (the study of the genome, i.e. the DNA and the regulation mechanisms at the level of genes), *transcriptomics* (the study of the RNA forming the transcriptome and the regulation activity therein), *proteomics* (the study of the proteins and their interactions with all other bio-molecules), is producing a huge amount of data concerning the behaviour of the single constituents of living organisms. With the help of *bioinformatics*, organizing and analyzing such data, we are assisting to an increase of understanding of the functional behaviour of such constituents. Nevertheless, this is not sufficient to gain a deep comprehension of how such components interact together at the system level, generating the set of complex behaviours we observe in life. This is the main motivation of the rising of systems biology, a new science integrating experimental activity and mathematical modeling in order to study the organization principles and the dynamical behavior of biological systems (see [142] for an overview of the field).

Mathematical and computational techniques are central in this approach to biology, as they provide the capability of describing formally living systems and studying their properties. Clearly, a correct choice of the abstraction level of such description is fundamental in order to grasp the key principles without getting lost in the details. Good introduction to the modeling art in systems biology are the books of Kitano [141] and Bower and Bolouri [39].

In the rest of the chapter, we give a brief and necessarily incomplete introduction to the field of mathematical and computational modeling in biology. Section 3.1 presents briefly the biological systems of interest, trying to underline the aspects deemed to be essential to understand their dynamics. Section 3.2 presents modeling techniques where the biological entities are represented as discrete quantities. Most of these approaches are based on a stochastic ingredient in order to introduce quantitative information in the models. The rationale behind this choice is presented in Sections 3.1.3 and 3.1.4, together with the famous Gillespie algorithm. Among these techniques, we are mainly interested in Stochastic Process Algebras and we present different formalisms that have been used in systems biology, namely Stochastic Petri Nets (Section 3.2.1), Stochastic π-calculus (Section 3.2.2), PEPA(Section 3.2.3) and the κ-calculus (Section 3.2.4).

Finally, Section 3.3 deals with modeling techniques based on a continuous representation of substances involved. Basically, these approaches are all based on differential equations, though different formats can be used. In Section 3.1.3 we present the main format used for modeling biochemical reactions, namely Mass Action equations, while in Section 3.3.3 we present an enhanced formalism, i.e. S-Systems (and Generalized Mass Action equations). Sections 3.3.1 and 3.3.2 deal with equations describing more sophisticated chemical kinetics than that of Mass Action, while Section 3.3.4 present an approach based on a Stochastic Differential Equation, the Chemical Langevin Equation, that is in the middle between deterministic and stochastic mass action kinetics.

3.1 Biological Systems in a Nutshell

In this section we give a brief overview of the biological systems we aim to model. This looks like an hard task, as even the simplest living being, i.e. a cell, is amazingly complicated, and so are its functioning mechanisms. We follow, however, the point of view that Luca Cardelli put forward in [47]. The main idea is to see biological activity inside cells as a form of *information processing*. This is not so strange, as a cell must process the information of the surrounding environment in order to feed and survive in it. However, any cellular process can be interpreted in this way, from duplication of DNA to regulation of gene expression, from signal transduction to membrane activity. Once we accept that biological systems are information processing machines, we can try to *abstract from chemical and physical details* in order to capture the *flow of information* that is intrinsic in their dynamics. This approach led Cardelli to identify some *abstract machines* associated to different machineries working in a cell. Each abstract machine tries to capture all the mechanisms that are sufficient to explain the systemic behaviour observed, describing the biological entities as objects (or better, *processes*) with a *discrete number of internal states* that can *change through interactions*.

Excluding small molecules, like water and minerals, there are essentially four classes of molecules that interact inside cells.

Nucleic Acids. There are five *nucleotides* (adenine, cytosine, guanine, thymine, and uracil) that form long polymeric chains: DNA and RNA. DNA stores the genetic information, while RNA participates in the process of extraction of this information, being the bridge molecule in the mechanism of construction of proteins from DNA.

Proteins. Proteins are sequences made up from about 20 different *amino acids*, that fold up in a peculiar and (more or less) stable 3D shape. Their functioning is determined mainly by a set of *active sites in their surface*, that can be active or inactive, depending on the interactions the protein is involved into. More information about their structure and the folding process can be found in Chapter 4.

Lipids. Lipids have mainly a structural function. In particular, cellular *membranes* are constituted by a double layer of phospholipids, whose structure gives to membranes unique properties, like the ability to self-assemble, to merge one into the other, and to contain interface proteins in their surface.

Carbohydrates. These are sugars, involved mainly in the energetic cycle of the cell. They are often linked into complex structures, some of them having with tree-like shapes.

Each of these classes of molecules is the basis of a different abstract machine. Proteins interact in complex *networks of biochemical reactions*, and are also involved in the regulation mechanism of DNA expression (*genetic regulatory networks*). Membranes, instead, are essential in the *transport networks*. Sugars, finally, are involved in metabolic processes and storage activity. In the rest of the thesis, we focalize our attention only to biochemical reactions and genetic networks, presenting them in deeper detail in Sections 3.1.1 and 3.1.2.

A common feature of these abstract machines is that the interactions among the substances involved form *complex networks*, whose structure is one of the main responsible of the cellular dynamics. These networks show an highly *non-linear behaviour*, induced by the presence of several *feedback* and *feed-forward* loops. Another property of these networks is *redundancy*; for instance, many genes can encode the same protein, though they are expressed in different conditions. Other characterizing features are *modularity* and *structural stability*. All these properties are responsible, for instance, of the amazing *robustness* exhibited by living beings in resisting and adapting to perturbations of the environment (see [142]).

3.1.1 Biochemical Reactions Networks

Biochemical reactions are chemical reactions involving mainly proteins. As in a cell there are several thousands of different proteins, the number of different reactions that can happen concurrently is very high. Depicting a diagram of such interaction capabilities, we obtain a complex network having all the characteristics mentioned at the end of the previous section.

Proteins are complex molecules by themselves, being constituted by a long chain of amino acids folded up in a peculiar 3D shape, see Chapter 4. This shape exposes in its surface a certain number of *interaction sites*, and those sites are precisely determining the functioning of the protein and its interaction capabilities. As we are interested in describing the effect of the interconnections and the resulting dynamics, we must understand what are the essential features we have to include in our models in order to capture the behaviour of the system. In fact, we cannot model everything explicitly, using, for instance, quantum mechanical considerations, as we would end up soon with intractable, hence useless, models. On the other hand, finding the right level of abstraction is probably the most difficult task in the art of modeling.

Forgetting interaction with small molecules, proteins are essentially involved in two kind of reactions. Some of their interaction sites have complementary shapes and physical/chemical properties, hence they can sting together like a velcro, resulting in the formation of a protein *complex*. The operation of *complexation* can modify the structure of a protein, exposing interaction sites that were previously hidden, or hiding some visible ones. The other possibility is that proteins modify the structure of each other by the addition or removal of a phosphate group (or other chemical groups, like a methyl group) at specific sites. These operations, similar in spirit to a boolean switching, are called *phosphorylation* and *dephosphorylation* (or, for instance, *methylation*, if the added group is a methyl). Proteins that phosphorylate another protein belong to the family of *kinase*, while dephosphorylation proteins are called *phosphatase*.

It turns out that the above operations do not depend on the fine details of the physical and chemical properties, hence proteins can be modeled as entities having a *finite number*

$$S_1 + E_1 \rightarrow S_2 + E_1$$
$$S_2 + E_2 \rightarrow S_1 + E_2$$

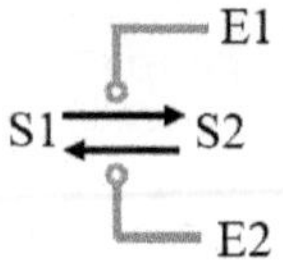

Figure 3.1: Model of a reversible reaction transforming substance S_1 into substance S_2, catalyzed in the two directions by two enzymes, E_1 and E_2. The reaction is described using both chemical equations (left) and a Kohn's map (right). The single headed arrows in the diagram represent substance transformation, while the circled arrow is associated to enzymatic activity.

of discrete states (automata), corresponding to the different interaction sites available, together with a list of interactions available in each state (and their strength, see Section 3.1.3).

The most widespread notation to represent biochemical networks is as a set of *chemical transformation rules*, which we call for simplicity *chemical equations*, of the form

$$m_1 R_1 + \ldots + m_k R_k \rightarrow n_1 P_1 + \ldots + n_h P_h,$$

where R_i are the *reactants*, P_i are the *products*, m_i and n_i are the *stoichiometric coefficients*. The equation above states that combining the reagents in the appropriate number we can obtain the products. Usually such equation is accompanied by a real number representing its basic expected "frequency". This number is related to the kinetic model adopted, see Section 3.1.3 for further information. A different way to represent biochemical reactions uses a graphical notation. One of the most renown is the one developed by Kohn [145], using different kind of arrows and having an implicit representation of complexes. In Figure 3.1 we show the description by both chemical equations and a Kohn map of a simple reversible reaction, catalyzed in the two directions by two different enzymes. *Enzymes* are proteins that are not modified by a reaction, but stimulate it. We present the enzymatic mechanisms in more detail in Section 3.3.1, where we discuss about Michaelis-Menten differential equations. Note that, in each representation of Figure 3.1, we are abstracting from the details of the enzymatic activity.

3.1.2   Genetic Regulatory Networks

The *central dogma* in molecular biology states that DNA is *transcribed* into RNA, which is then *translated* into a protein. The flow of information, however, goes also in the other direction: proteins interact with the DNA transcription mechanism in order to regulate it. DNA is composed by several (thousand of) *genes*, each encoding a different protein. Genes are roughly composed by two regions: the regulatory region and the coding one. The *coding region* stores the information needed to build the coded protein using the *genetic code* to associate triples of nucleotides to aminoacids. The *regulatory region*, usually found upstream of the coding part of the gene and called *promoter*, is involved in the mechanisms that control the expression of the gene itself. The regulation is performed by dedicated proteins, called *transcription factors*, that bind to specific small sequences of the promoter, called *transcription factors binding sites*. There are basically two different kinds of transcription factors, the *enhancers*, increasing or enabling the expression of the coded protein, and the *repressors*, having a negative control function. Transcription factors act on DNA in complex ways, often combining in complexes before the binding or during it. Also binding sites, like interaction sites on proteins, can be exposed or covered, both by the effect of a bound factor or by conformational properties of the DNA strand, though this last form of regulation is not well understood yet.

An abstract representation of genes is as *logical gates* of a rather peculiar type: they have a fixed output, the produced protein (though the rate of production can vary due to the effect of transcription factors), and a variable number of inputs, corresponding to different binding sites. Each input has an affinity that describes the strength of interaction of the transcription factor (that can be seen as the expected frequency of interaction per unit of factor). This description can be visualized in a wired diagram. This diagram is rather complex, as it has to encode all the combinatorial regulations happening in the promoter of each gene.

The complexity of genetic regulatory mechanism is not only in the complexity of the networks describing it, but also in the fact that transcription factors are proteins, hence they can undergo a series of biochemical modifications depending on several factors in the cell. For instance, a transcription factor can be activated only by a cascade of events triggered by an "input" signal outside the cell. Therefore, a cell can adapt to environmental modification by changing the profile of expression of its genes. As a consequence, genetic regulatory and biochemical networks are not independent subsystems, but rather they are strictly interconnected. However, a big difference between the two is in the *time scale*: while biochemical reactions take time in the order of milliseconds, the transcription of a gene can take minutes. This fact poses several challenges in the integration of the two models, due to the high degree of *stiffness* of the resulting system.

3.1.3   Chemical Kinetics

Chemical (and biochemical) kinetics is concerned with time-evolution of a chemical mixture whose substances react according to a specific set of chemical reactions. In particular, the interest is mainly in the behaviour of the system away from equilibrium. To specify the full dynamical behaviour of a set of equations, we need to know also the rate at which each reaction occurs. In order to make clear the concept of rate, we need to fix a formal setting where to model kinetics: either continuous and deterministic or discrete and stochastic. As we will see in the following, the discrete and stochastic description is physically sounder.

Mass Action Deterministic Kinetics

In the classical setting, the amount of a chemical substance is generally regarded as a concentration, thus expressed in units like moles per liter. This is probably a heritage of the experimental study of chemical reactions in "beakers of water". Conventionally, the concentration of a chemical species X is denoted $[X]$.

The *law of mass action kinetics* states that *the rate of a chemical reaction is directly proportional to the product of the effective concentrations of each participating molecule.* Basically, it is proportional to the concentration of reactants involved raised to the power of their stoichiometry.

For example, consider the following set of reactions, known also as the Lotka-Volterra system (cf. [102]):

$$
\begin{aligned}
A &\rightarrow 2A \\
A + B &\rightarrow 2B \\
B &\rightarrow \emptyset.
\end{aligned}
\tag{3.1}
$$

This system of equations aims to model a simple prey-predator dynamics, where A is the prey and B is the predator. Here, preys reproduce "out of nothing" (unlimited food reserve), as stated by the first equation, while predators reproduce while eating preys and can also die.

The rate of the second reaction is, according to the mass-action law, proportional to the products of the two concentrations of the reactants involved, i.e. it is proportional to $[A][B]$. Indicating with k_2 the constant of proportionality, we have that this rate is $k_2[A][B]$. $k_2[A][B]$ is usually known as the *rate law*, while k_2 is known as the *rate constant*.

If we consider all the reactions involved, we can write a system of differential equations giving the speed of change of the concentration of each molecule. To write the equation for X, we simply have to sum the rate law terms of the different equations involving X, multiplying by $+1$ all the terms coming from equations where X is a product and by -1 the terms corresponding to equations where X is a reactant.

For the Lotka-Volterra set of reactions, the corresponding equations are

$$
\begin{aligned}
\frac{d[A]}{dt} &= k_1[A] - k_2[A][B] \\
\frac{d[B]}{dt} &= k_2[A][B] - k_3[B].
\end{aligned}
\tag{3.2}
$$

Once we have this set of equations, we can analyze it. For instance, we can solve it numerically for given initial conditions, we can study its equilibrium points, or we can analyze its dependence on parameters, looking for bifurcation points and chaotic regions. See Figure 3.2 for a visual example, or books on dynamical systems, like [227].

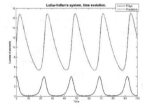 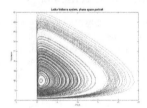

Figure 3.2: Lotka-Volterra system. (**left**) Solution of the mass action equations (3.2) for $k_1 = 1$, $k_2 = 0.1$, $k_3 = 0.1$, with initial conditions $A_0 = 4$ and $B_0 = 10$. (**right**) Phase portrait of the Lotka Volterra system, for the same value of parameters as above. As we can see, all the solutions show an oscillating behaviour. The system has an (instable) equilibrium for $A = 1$, $B = 10$, at the center of the circles.

Mass Action Stochastic Kinetics

The continuous description of molecules as concentrations fails to capture the discrete and stochastic nature of chemical systems. Indeed, these features are particulary relevant when the concentration of some reagents are very low (or, staten in another way, when the number of molecules of a certain kind is small). In these cases, the stochastic fluctuations exhibited by all molecular systems can have a dramatic impact on the behaviour of the system, as shown, for instance, in [234], for a simplified version of the molecular clock governing circadian rhythm.

Why chemical systems are stochastic? A simple model of a such a system is as a container where all the molecules are in an homogeneous mixture. Due to thermic agitation, all the molecules move in a random manner, essentially driven by Brownian motion. Therefore, if we abstract from the exact position and velocity of each single molecule, we are left with a system where the position of each molecule is represented by a random variable. Thus, chemical systems are actually stochastic when described at the level of granularity of number of molecules of each species (or concentrations).

We want now to understand the stochastic ingredient in deeper detail, following predominantly the presentation given in [239]. In particular, it happens that the rate at which two molecules collide is constant whenever temperature and volume are fixed over time, if the system is in thermal equilibrium, or well stirred. This fact can be demonstrated using statistical mechanic arguments, see the article of Gillespie [100] for further details. The essential argument here is that molecules are uniformly distributed throughout the volume V, and this distribution is independent of time. This means that the probability that a molecule is in a region D of volume V' is proportional to this volume, i.e., denoting with P the random variable identifying the position of the molecule,

$$\mathbb{P}(P \in D) = \frac{V'}{V}.$$

If we want now to compute the probability of two molecules being within reacting distance r (with r very small, in order to ignore boundary conditions), we can condition on the position of one of them, and then compute the average of the conditional probability, resulting in:

$$\mathbb{P}(|P_1 - P_2| < r) = \mathbb{E}[\mathbb{P}(|P_1 - P_2| < r \mid P_2)].$$

The uniformity of P_2 implies that the expectation above is redundant, hence the probability of being at reacting distance equals the probability of one molecule being ia a sphere of radius r with given center p, call it $S(p, r)$:

$$\mathbb{P}(|P_1 - P_2| < r) = \mathbb{P}(P_1 \in S(p, r)) = \frac{4\pi r^3}{3V}.$$

Now, given that two molecules are within reaction distance, they will react with a probability independent of the the volume V (but dependent, for instance, on their chemical affinity), thus resulting in a rate of reaction inversely proportional to the volume V. If this volume is constant, as all other quantities like temperature and pressure, the rate of reaction c of two molecules is constant. In addition, two pairs of molecules can undergo the reaction independently with the same rate. This soon implies that the global rate of a reaction is proportional to the number of pairs of molecules present in the system, where the constant of proportionality is indeed the basic rate of reaction c.

To construct the stochastic model, given the expression of rates of reactions, we proceed in the following way. Suppose a system involves u different species $X_1, \ldots, X_u$ and v chemical reactions $R_1, \ldots, R_v$. Then, each reaction R_i has a *stochastic rate constant* c_i

and, indicating with $\mathbf{x}$ the u-vector of integers representing the number of molecules for each chemical species, a *rate law function* $h_i(\mathbf{x}, c_i)$. For a zeroth-order reaction $R_i : \emptyset \rightarrow^{c_i} X$, the corresponding rate law function is $h_i(\mathbf{x}, c_i) = c_i$; for a first order reaction $R_i : X_j \rightarrow^{c_i}?$ we have $h_i(\mathbf{x}, c_i) = c_i x_j$. The case of second order reactions is more complicated, as we have to distinguish between reactions having molecules of different species as reactants, like $R_i : X_j + X_h \rightarrow^{c_i}?$, and reactions involving two molecules of the same species, $R_i : 2X_j \rightarrow^{c_i}?$. In the first case we have $h_i(\mathbf{x}, c_i) = c_i x_j x_h$, while in the second $h_i(\mathbf{x}, c_i) = c_i \frac{x_j(x_j-1)}{2}$, as we need to count correctly the number of pairs of molecules.

The interpretation of the rate function h is that the probability that an R_i reaction happens in time $[t, t+dt]$ is $h_i(x, c_i)dt$. This means that, considering a system with only reaction R_i, the time to such reaction event is exponentially distributed with rate $h_i(x, c_i)$. As the time interval dt considered is infinitesimal, we can safely assume that only one reaction event occurs during it. In fact, the probability that two such events happen is of the order of dt^2, and can be safely ignored. Note that the function h_i depends only on the current state of the system, hence the resulting stochastic model is a Continuous Time Markov Chain. In fact, it can be proven [102] that the probability of observing a reaction R_j in the time interval $[t, t+dt]$ factorizes as

$$\mathbb{P}(R_j, t)dt = \frac{h_j(\mathbf{x}, t)}{h_0(\mathbf{x}, t)} h_0(\mathbf{x}, t) e^{-h_0(\mathbf{x},t)t} dt, \tag{3.3}$$

where

$$h_0(\mathbf{x}, t) = \sum_{i=1}^{v} h_i(\mathbf{x}, t). \tag{3.4}$$

Therefore, the probability of next reaction being R_j is independent on the time elapsed and proportional to the rate of the reaction itself, while the time is distributed exponentially with rate $h_0(\mathbf{x}, t)$. This corresponds to the characterization of a Markov Chain by jump chain and holding times. The forward equation for this Markov Chain (see Section 2.3 of Chapter 2 and Section 3.3.4 hereafter) is often referred in literature with the name of *Chemical Master Equation*.

As a final remark, we observe that the constants for stochastic simulation and for mass action differential equations are not the same, but differ from a factor connecting them to the volume. For instance, for a zero order reaction, the stochastic rate is V times the deterministic rate, $c = Vk$. For a first order reaction it holds that $c = k$, while for a second order one the relation is either $c = k/V$, if reactants belong to two different species, or $c = 2k/V$, if the reaction involves two molecules of the same kind.

3.1.4   Gillespie Algorithm

The discussion in the previous section shows that the stochastic process representing the temporal evolution of a chemical system, under the hypothesis of constant volume and temperature and of homogeneity, is a Continuous Time Markov Chain with discrete state space, where each state is a tuple of integers representing the number of the different molecules. Equation (3.3) is at the heart of the *Gillespie's algorithm* (or *Gillespie's direct method*) [101, 102], which is the standard procedure to simulate such systems. In fact, such algorithm is a Monte Carlo simulation of the Markov Chain, based on its characterization with jump chain and holding times (that are recovered from Equation (3.3)): the time to next reaction is drawn from an exponentially distributed random variable with rate $h_0(\mathbf{x}, t)$, while the reaction type is drawn independently with probability distribution $h_i(\mathbf{x}, t)/h_0(\mathbf{x}, t)$ (h_0 is defined by Equation (3.4)).

At each step, the algorithm works by drawing independently two uniform random numbers in $(0, 1)$ (using a pseudo-random generator), and then by applying standard Monte Carlo methods (see Section 2.4) in order to draw the time elapsed and the next reaction according to probability distributions described above. After that, the state of the system is updated by the prescriptions of the chosen reaction and by increasing the global time. Usually, the algorithm stops after reaching an user-defined maximum time.

3.2 Discrete Modeling Techniques

In this section we present some modeling approaches to biological systems, that have in common the discrete representation of entities into play. The literature on modeling and analysis of biological systems is huge, so we do not even try to give an overview on it, referring the interested reader to [70, 39, 141]. Rather, we present some techniques that, in one way or another, will have a role in the work hereafter.

In the class of discrete approaches to model biological systems, an important position is occupied by techniques coming from the field of concurrent software systems. In fact, these theories were developed to represent and analyze properties of large systems of computational entities acting in parallel, communicating and competing for shared resources. As argued in the introduction, and for instance in [203], there is a striking parallelism between these systems and the biological ones. In particular, stochastic Petri Nets [110] and Stochastic Process Algebras [195, 44], theories born in the field of performance analysis of computing systems, have been used extensively in biological modeling. In the following, we present briefly the use of Stochastic Petri Nets (Section 3.2.1), Stochastic π-calculus (Section 3.2.2) and PEPA (Section 3.2.3) for such task. As a final technique, we present the κ-calculus (Section 3.2.4), a process algebra that has been designed explicitly to deal with the basic operations of the protein machine, like complexation and (de)phosphorylation.

3.2.1 Stochastic Petri Nets

Stochastic Petri Nets [117] (SPN) are an extension in the stochastic realm of Petri Nets [190], a formalism used for describing concurrent systems, dating back to sixties. SPN have been used recently to model biological systems of various kinds, mainly biochemical reactions, see [110, 191, 118] for further details. SPN can be seen as a *bipartite graph*, having two different kind of nodes: *places* and *transitions*. Being bipartite, arcs can go just from places to transitions and from transition to places. Intuitively, places represent computational locations, while transitions are the action capabilities of the system. Each place contains a certain number of tokens, representing the resources available at that location. The vector storing the number of tokens for each position is called the *marking* of the Petri Net. Each arc of the net is labeled with an integer, representing the number of tokens needed or produced by a particular transition. A transition is enabled if in each of the places having edges entering the transition node (let's call them *reactants*) there are at least as many tokens as required by the labels of edges. An enabled transition

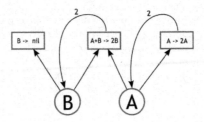

Figure 3.3: Stochastic Petri Net associated to the Lotka-Volterra chemical system, described by chemical equations (3.1). Places are represented by circles, while transitions corresponds to rectangles. Edges go only from places to transitions and viceversa, and they are labeled by integers, giving the stoichiometry of the corresponding reaction. Each transition has a basic rate associated to it, and its stochastic rate is computing using the h functions of Section 3.1.3.

can *fire*, removing the tokens from the entering places and producing tokens in the locations reachable crossing edges exiting from the transition (call them *products*). Tokens are produced with the multiplicity indicated by the labels of such edges. In SPN, each transition has a basic rate associated to it, that is multiplied by the product of tokens in reactants (raised to their stoichiometry) in order to compute the rate of the transition, given the current configuration of the system. At each instant, all active transitions compete for firing in such a way that the fastest one is executed. Then the system loses memory of the race condition, and it starts another competition anew. The resulting stochastic process is clearly a Continuous Time Markov Chain, with discrete state space.

There is a straightforward mapping from chemical reactions, described by chemical equations, to Stochastic Petri Nets. In fact, *each molecule of the system corresponds to a place and each reaction to a transition*. Edges connect reactants to transitions and transitions to products, and tokens represent the number of molecules of a certain kind present in the system.

As an example, in Figure 3.3, we show the SPN associated to the Lotka-Volterra system described in equations (3.1). As we can see, we have two places, corresponding to preys (A) and predators (B), and three transitions, corresponding to the three chemical equations of (3.1). If the basic rate of the transition in the middle is c_2, then the rate of the transition is given by $c_2 AB$, which is exactly the expression of the stochastic rate given by the h functions defined in Section 3.1.3.

3.2.2   Stochastic π-calculus

Process Algebras are algebraic theories whose aim is that of modeling the communication interactions among concurrent processes. Most of them are composed by two basic entities: agents and communications channels. Intuitively, agents are entities that can interact by exchanging messages through shared channels. All the focus here is concentrated on interactions, while all other details of computations are abstracted away. Despite this, the computational models arising are Turing-complete [212]. There is a huge plethora of

$$Program = D.P$$

$$D ::= \varepsilon \mid D.D \mid X ::= P$$

$$\pi ::= \overline{x}y \mid x(z) \mid \tau$$
$$P ::= M \mid P \parallel P' \mid X$$
$$M ::= \mathbf{0} \mid \pi.P \mid M + M'$$

Table 3.1: Syntax of restricted π-calculus

different process algebras, differing in the communication style, in the the syntax, in the semantics, and so on. Among them, we recall CCS [172] and π-calculus [174, 212], that have a synchronous handshaking-based communication style.

The introduction of the stochastic ingredient, rendering quantitative information in the model, is achieved associating real numbers, i.e. rates, to channels. Intuitively, each transmission along a channel has a duration, represented by an exponentially distributed random variable [181], whose rate is the real number labeling the channel.

The quantitative information attached to channels induces a continuous-time stochastic evolution of the system. In each configuration, there is a race among all active transitions (i.e. all possible communications that can happen), and the fastest one is executed. Thus, the semantic structure associated to stochastic π-calculus [194] is a Continuous Time Markov Chain (Section 2.3), whose states are configurations of the program and whose transition's rates are determined by the basic rates of the associated channels, multiplied by the number of pairs of agents ready to communicate on these channels.

An execution trace of a stochastic π-calculus program can be computed using the Gillespie's method, see Section 3.1.4, that has been used as the basic simulation engine to implement SPiM [51], a popular simulator for stochastic π-calculus.

Stochastic π-calculus has been used as a modeling technique for biochemical reactions in [197], where the interesting idea of a parallel between biological molecules that interact and agents that communicate has been exploited. Since then, several biological systems have been modeled [195, 20], showing the feasibility of the technique. One of its main appeals resides in the *compositional* nature of the modeling process, allowing to describe single biological entities with a concern on of the logic of their interactions only, while complex behaviors emerge as a system's property of the entire model.

In the following we introduce the syntax of a simplified version of π-calculus, where we forget some operators like restriction and the creation of fresh channel names. This choice is mainly determined by the need of keeping the presentation simple, and by the fact that this restricted version, called by Cardelli chemical π-calculus in [48], suffices to describe networks of biochemical reactions and genetic networks. The restricted π-calculus is presented in Table 3.1. The two basic actions are those devoted to send and receive names along a channel. In addition, we have a choice operator, allowing to select one among different alternatives, and a parallel composition, enabling agents to be executed together. Recursion is achieved through the use of constants within the syntactic specification of agents. As said above, each channel name has a basic stochastic rate attached to it.

As an example, we give the stochastic π-calculus model of the Lotka-Volterra chemical system, described by equations (3.1). The idea in modeling with π-calculus is to associate a process to each entity of the system. For instance, each pray and each predator is described by a π-process. The number of processes of each kind corresponds to the amount of each species in the system. Therefore, if we have 100 preys, then we will have 100 prey processes in parallel. The behavior of a single prey can be described easily: either it reproduces, thanks to the food in the environment, or it interacts with a predator, being eaten. Both actions happen at a certain basic rate, the one specified in the chemical equations. The π-process for a prey is therefore:

$$A \ ::= \ \tau_{c_1}.(A \parallel A) + e().\mathbf{0}. \tag{3.5}$$

Thus, the unspecified interaction with the environment is modeled as a τ (silent) action, after which the process forks generating two new prey processes. The interaction with a predator, instead, is obtained by a synchronization along channel e (for eating). In this case, the prey receives a message along his channel, which need to be sent by a predator process, of the form

$$B \ ::= \ \tau_{c_3}.\mathbf{0} + \overline{e}().(B \parallel B) \tag{3.6}$$

The predator can either die (silent action) or eat a prey (sending a message along channel e) and duplicate (by forking as $B \parallel B$). Note that both τ actions and the channel e have a stochastic rate attached to it. The semantics of stochastic π-calculus guarantees that the global rate of the actions of the system equals the rate specified by the function h, see [194] for further details.

3.2.3   PEPA

PEPA (Performance Evaluation Process Algebra) [127, 128] is a process algebra designed to tackle performance related problems. Therefore, it has been provided with a stochastic semantics since the beginning (unlike π-calculus).

The basic elements of PEPA are actions (α, r), identified by a name α and a rate r, being the parameter of an exponential distribution associated to their duration. Actions can be executed sequentially using the prefix operator ".", like in $(\alpha, r).P$. Then we have the summation operator $P_1 + P_2$ to model stochastic choice, the name definition $X = E$, modeling recursion, the hiding operator P/L, where L is the set of actions that is made local to P. Finally, the last operator is the cooperation $P_1 \overset{L}{\bowtie} P_2$, requiring processes P_1 and P_2 to synchronize on the actions of L. This operator is associative and commutative for a fixed set L, and synchronization in $P_1 \overset{L}{\bowtie} \ldots \overset{L}{\bowtie} P_n$ must involve all n processes, meaning that actions in L cannot be executed unless *all* processes are ready to perform one of them. Processes can be synchronized on the empty set, meaning that they can evolve independently; the operator $\overset{\emptyset}{\bowtie}$ is usually denoted by $\parallel$. When actions are synchronized, the rate of the transition is usually chosen as the minimum of all the rates involved. Actions can also have attached a passive rate, denoted by $\top$, which does not influence the computation of the synchronizing rate. All actions with passive rates are required to be under the scope of a cooperation operator.

Modeling biological systems in PEPA [44] is more or less similar in spirit to π-calculus: one models each molecule of the system, and then makes them cooperate. A little care must be put in the combination of processes through the cooperation operator, as operators defined with respect to different set of actions do not commute. For example, a model of the Lotka-Volterra system (equations (3.1)) is given by

$$A = (reproduce, c_1).(A \parallel A) + (eating, \top).\mathbf{0}$$
$$B = (die, c_3).\mathbf{0} + (eating, c_2).(B \parallel B)$$

$$Model = (A \parallel \ldots \parallel A) \overset{\{eating\}}{\bowtie} (B \parallel \ldots \parallel B)$$

(3.7)

However, PEPA usually comes with a syntactic restriction to the language, allowing to compose in parallel only *sequential processes*, i.e. processes that are defined without using the coordination operator. This basically means that the number of agents in parallel is always finite and fixed during a computation. The rationale of this restriction is that of guaranteeing finiteness of the underlying CTMC, in order to perform computational analysis of it, like computing the steady state distribution, cf. [128]. This modification, however, rules out the model shown in equation (3.7). A possible solution is presented in Table 3.2; the model is explained in the caption.

Recently, it has been developed bioPEPA [53], a new version of PEPA specifically tailored to model biochemical reactions. Among its features, we remark particularly the introduction of functional rates, allowing to describe more general chemical kinetics than mass action (see Section 3.1.1 for a discussion on this subject).

Comparing the modeling activity of PEPA and π-calculus, we can see that the main difference regards the synchronization mechanism: in PEPA it may involve n processes, while in π-calculus it is always binary. Note that this may be not a limitation, at least for chemical reactions, that are at most binary in nature. In addition, PEPA models have always a finite state space, permitting the use of the broad range of analysis techniques present in the market. The negative byproduct, however, is a loss of elegance.

3.2.4 κ-calculus

In Section 3.1.1 we argued, following the lines of [47], that in models of biochemical networks proteins can be identified with their interface, composed by interaction sites and boolean switches (that are triggered by (de)phosphorylation). Proteins can combine in complexes, in rather intricate ways. For instance, if a protein has four interaction sites, then usually different proteins can bind at different sites, in variable order and independently from the status of other sites, giving rise to a big number of admissible complexes. In addition, the creation of certain bounds can hide some interaction sites if they are not occupied. For instance, consider the simple Kohn map of Figure 3.4[1]. There, protein A can bind both with protein B and C, so all intermediate complexes AB, ABC, AC can be formed. However, if A is bound with B, then the binding with C is inhibited, while A can always bind B, despite the presence or absence of a bound with C.

Modeling this complex behaviour using a simple process algebra like π-calculus can be an hard activity, as these interdependencies complicate heavily the model. In [65], the authors propose a calculus that has *complexation* and *activation* as basic primitives. Proteins are modeled by an identification name and by two multiset of sites, the first containing *visible sites* and the second containing *hidden sites*. Formally, this is denoted

---

[1]Kohn maps will be presented in more detail in Section 7.4.1.

$$A = (reproduce, c_1).A + (eating, \top).A'$$
$$A' = (resurrect, \top).A$$

$$A_r = (reproduce, \top).A'_r$$
$$A'_r = (resurrect, c_{fast}).A_r$$

$$Prey = (\underbrace{A \parallel \ldots \parallel A}_{n_1} \parallel \underbrace{A' \parallel \ldots \parallel A'}_{n_2})^{\overset{\{resurrect, reproduce\}}{\bowtie}} (\underbrace{A_r \parallel \ldots \parallel A_r}_{n_1+n_2})$$

$$B = (die.c_3).B' + (eating, c_2).(reproduce, c_{fast}).B$$
$$B' = (resurrect, \top).B$$

$$B_r = (reproduce, \top).B'_r$$
$$B'_r = (resurrect, c_{fast}).B_r$$

$$Predator = (\underbrace{B \parallel \ldots \parallel B}_{m_1} \parallel \underbrace{B' \parallel \ldots \parallel B'}_{m_2})^{\overset{\{resurrect, reproduce\}}{\bowtie}} (\underbrace{B_r \parallel \ldots \parallel B_r}_{m_1+m_2})$$

$$Model = Prey \overset{\{eating\}}{\bowtie} Predator$$

Table 3.2: Model of the Lotka-Volterra system, described by equations (3.1). The model is written in PEPA, with the restriction of having only sequential agents composed in parallel. Therefore, the number of parallel agents is constant during each computation, leading to the problem of managing birth and death of prey and predators. The idea is to put in parallel at the beginning of the computation a fixed number of processes, in two different states, alive and dead. Hence, there is the need of a mechanism to resurrect dead processes when a new offspring is generated. The synchronization combinator of PEPA works on all processes under its scope, so we need to introduce "resurrector" agents that synchronize on just one dead agent. This is achieved by putting in parallel all dead and alive agents with no synchronization, and then synchronizing this composition with some resurrecting agents, again in parallel, see the definition of agent *Prey* or *Predator*. The synchronization is performed on the actions *resurrect* and *reproduce*; synchronization on *reproduce* activates a resurrect agent, while synchronization on *resurrect* changes a dead agent into an alive one. The resurrection operations are not part of the original model, so we must be careful in assigning their stochastic rate, in order not to alter the behaviour of the system. A solution is simply that of using for these operations a rate which is several order of magnitude bigger than rates of the original system, resulting in actions that can be seen as instantaneous. After defining *Prey* and *Predator* agents, we synchronize them on the *eating* action in order to build the entire model.

Figure 3.4: Kohn map representing the formation of a complex between three proteins, A, B, and C. The complexes are represented by the circle in the middle of the double arrowed edges. The red arrow with a bar at one end indicates inhibition, meaning that the formation of the complex AB inhibits the binding of C to A. Note that these maps represent complexes implicitly.

by $A(\rho, \sigma)$, where ρ is the multiset of hidden sites and σ is the multiset of visible sites. Sites are simply identified by names.

The activation operation has the effect of moving some sites from σ to ρ or viceversa. It may involve one or more proteins (usually at least two). Complexation, instead, is used to form protein complexes of the form $A_1(\rho_1, \sigma_1) \odot \ldots \odot A_n(\rho_n, \sigma_n)$. It is intended that each binding consumes a pair of interaction sites, one for each protein, and these bound sites are removed from the sets σ. Both operations are local and can be embedded in bigger complexes. Both operations are described by reaction rules, while proteins and complexes are combined together using the solution operator ",". An interesting feature of κ-calculus is that it allows an almost direct translation from Kohn maps, and the authors provide also a mapping from κ-calculus to π-calculus. Therefore, this calculus can be seen as an intermediate layer between biological notation and computational one.

Since the core version of [65], κ-calculus has been extended and refined. More recent versions [62, 64], implement complexation by labeling connected sites with a unique label, instead of removing them, introduce a more general notion of state of a site as a valuation, and introduce a stochastic model by assigning rates to reactions. The obtained calculus can be seen as a grammar describing graph rewriting rules, thus equating connected graph with complexes and collections of connected graphs with solutions. The analysis performed by the authors on their models are not limited to stochastic simulation, but they also try to reconstruct meaningful pathways by analyzing conflict and concurrency of events and reconstructing causal histories of computations [64].

All in all, κ-calculus describes biological systems in a rule-based fashion (similarly to the work of [86]), an approach which, in our opinion, reveals to be more natural and closer to biologist's way of thinking.

3.3 Continuous Modeling Techniques

In this section we deal with modeling techniques using a continuous representation of the entities into play. As our main concern is in the dynamics of the system, all these modeling approaches are based on the mathematical instrument of differential equations, either deterministic or stochastic. In Section 3.1.3, we have already seen a modeling formalism for biochemical reactions based on differential equations. These equations have a fixed format, i.e. their addends are products of variable concentration, possibly raised to an integer exponent, multiplied by a constant, the constant rate of the reaction. These equations generally go under the name of *mass action equations*.

First, we present two classes of equations that are used to describe biochemical reactions at an higher level of abstraction than mass action ones. Michaelis-Menten equations (Section 3.3.1) deal with single enzymatic reactions, while Hill's equations (Section 3.3.2) model a sequence of enzymatic reactions having a cooperative effect.

Next, we present a general format of ODE, the S-System (Section 3.3.3), that is more expressive than mass action, and it is able to capture some aspect of biological modeling, like positive and negative regulation, in a consistent framework.

Finally, we present an approach based on a stochastic differential equation called Chemical Langevin Equation [98] (Section 3.3.4) that is somehow in the middle between deterministic and stochastic mass action kinetics.

3.3.1 Michaelis-Menten Equations

An enzymatic reaction is a chemical reaction that transforms a substrate S into a product P by means of the action of a protein that is not transformed in the process, called *enzyme*. An enzymatic reaction can be described by the following set of chemical equations:

$$
\begin{aligned}
E + S &\to_{k_1} ES \\
ES &\to_{k_{-1}} E + S \\
ES &\to_{k_2} E + P
\end{aligned}
\tag{3.8}
$$

Essentially, the substrate S and the enzyme E can bind together forming an enzyme-substrate complex ES. This complex can both dissociate back into the enzyme and the substrate, or undergo some internal modifications and transform the substrate into the product P, releasing again the enzyme.

The model of this system using Mass Action equations is the following:

$$
\begin{aligned}
\frac{d[S]}{dt} &= k_{-1}[ES] - k_1[S][E] \\
\frac{d[E]}{dt} &= (k_{-1} + k_2)[ES] - k_1[S][E] \\
\frac{d[ES]}{dt} &= k_1[S][E] - (k_{-1} + k_2)[ES] \\
\frac{d[P]}{dt} &= k_2[ES]
\end{aligned}
\tag{3.9}
$$

In Figure 3.5 we can see a solution of these equations: the substrate is converted into the product, and, during this process, all the enzyme is in the form of the complex ES.

This last observation suggest that, during the transformation of S into P, the derivative of ES is approximatively zero. Whenever this is the case, we say that the system satisfies the *quasi-steady state assumption*, i.e. it hold that $\frac{d[ES]}{dt} = 0$. This happens usually if the concentration of the substrate is much higher than the concentration of the enzyme, and if the total concentration of the enzyme (both in free and in complex form) remains constant.

Under quasi-steady state hypothesis, we can derive a particular format of equations describing essentially the same dynamics of (3.9). From the relation $\frac{d[ES]}{dt} = 0$, i.e. $k_1[S][E] - (k_{-1} + k_2)[ES] = 0$, we derive $[ES] = \frac{k_1[S][E]}{k_{-1}+k_2}$, which, calling $K = \frac{k_{-1}+k_2}{k_1}$, becomes

$$
[ES] = \frac{[S][E]}{K}.
$$

The constant K is usually referred to as the *Michaelis-Menten constant* of the enzymatic reaction.

The hypothesis of a constant quantity of E can be written as $[E_0] = [E] + [ES]$, where $[E_0]$ is the initial quantity of the enzyme E. This can be rewritten as $[E] = [E_0] - [ES]$; putting this relation into that for $[ES]$, we obtain, after simple algebraic manipulations, that

$$[ES] = \frac{[E_0][S]}{K + [S]}.$$

We can now substitute this expression for $[ES]$ in the differential equation for P, getting

$$\frac{d[P]}{dt} = \frac{k_2[E_0][S]}{K + [S]} = \frac{V_{max}[S]}{K + [S]} \tag{3.10}$$

This equation goes under the name of *Michaelis-Menten kinetic law*. V_{max} is defined as $V_{max} = k_2[E_0]$, and it represents the maximum velocity of transformation of S into P (in fact V_{max} is the limit velocity for $[S]$ going to infinity). The constant K can be measured easily, as it equals the concentration of S when the velocity is $\frac{1}{2}V_{max}$.

Figure 3.5: Dynamics of an enzymatic reaction. (**left**) Mass Action dynamics, with parameters $k_1 = 1$, $k_{-1} = 0.1$, $k_2 = 1$, and initial conditions $S = 50$, $E = 10$, $ES = P = 0$. (**right**) Michaelis-Menten dynamics, with parameters $V_{max} = 10$, $K = 1.01$ and initial conditions $S = 50$, $P = 0$.

In Figure 3.5, we can see the time evolution of the Michaelis-Menten equation, compared with the evolution of the corresponding Mass actions.[2] We can see that the behaviour is essentially the same (in this case, the quasi-steady state assumption is fulfilled, as the trace of $[E]$ and $[ES]$ show).

3.3.2 Hill's Equation

Sometimes enzymes manifest a form of cooperativity while binding to substrate. In fact, if the enzyme has more than one binding site, the interactions between binding sites can induce a form of cooperativity. If the binding of ligand at one site increases the affinity for ligand at another site, the macromolecule exhibits positive cooperativity. Conversely, if the binding of ligand at one site lowers the affinity for ligand at another site, the protein exhibits negative cooperativity. If the ligand binds at each site independently, the binding is non-cooperative, see Figure 3.6 for a visual summary.

---

[2]The correspondence between the two systems is obtained by calculating Michaelis-Menten parameters starting from Mass Action rates.

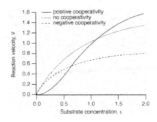

Figure 3.6: Cooperative effect of interaction between different binding sites in an enzyme.

For example, if the enzyme E has two binding sites, both interacting with substrate S, we have the following system of reactions:

$$E + S \underset{k_{-1}}{\overset{k_1}{\rightleftarrows}} C_1 \rightarrow_{k_2} E + P$$
$$C_1 + S \underset{k_{-3}}{\overset{k_3}{\rightleftarrows}} C_2 \rightarrow_{k_4} E + P \tag{3.11}$$

If we impose the quasi-steady state assumption also in this case, meaning that the derivative of $[C_1]$ and $[C_2]$ is approximatively zero during the transformation of S into P, after some calculations (see [58] for further details), we obtain the following equation for the velocity of the reaction:

$$\frac{d[P]}{dt} = \frac{(k_2 K_2 + k_4[S])[E_0][S]}{K_1 K_2 + K_2[S] + [S]^2},$$

where K_i, $i = 1, 2$ are the Michaelis-Menten rates for the two reactions of (3.11). Note that the previous equation has a quadratic dependance from $[S]$ both in numerator and denominator. Under the (strong) assumption of $k_3 \rightarrow \infty$, $k_1 \rightarrow 0$, $k_1 k_3$ constant (basically, two molecules of the substrate bind at the same time), the previous equation boils down to

$$\frac{d[P]}{dt} = \frac{V_{max}[S]^2}{K_m^2 + [S]^2}, \tag{3.12}$$

which resembles the equation of Michaelis-Menten, with a quadratic exponent to $[S]$ and K.

If we generalize the previous reasoning to the case of an enzyme binding n molecules of the substrate, we obtain the equation (3.13), called *Hill equation of degree n*. The exponent n, in particular, is called *Hill coefficient*.

$$\frac{d[P]}{dt} = \frac{V_{max}[S]^n}{K_m^n + [S]^n}, \tag{3.13}$$

The name Hill comes from Archibald Hill, an English physician that introduced such equations in 1910 as a model of the cooperative effect of oxygen binding in hemoglobin.

3.3.3   S-Systems

Modeling biological systems by means of ordinary differential equations (ODE) means to identify a law representing the evolution over time of the concentration of each of the biological entities under study. We have already seen a general purpose formalism, the mass action equations, stemming out directly from the principle of mass action of biochemical reactions, plus some specialized formalism to tackle enzymatic reactions. Mass Action equations are fine as long as biochemical reactions are specified in their full detail, tough this is not always the case. In fact, biologists often talk about stimulation or inhibition effects of proteins, without fully understanding the biochemical mechanisms behind. Consequently, we need a formalism that can work at an higher level of abstraction, representing these effects in an implicit way. In literature, there are several proposals [83], though here we focus mainly on two general purpose formats of equations: Generalized Mass Action equations (GMA) and S-Systems [236].

S-System equations (which are a special case of GMA equations) describe the evolution of a set of dependent variables $(X_1, \ldots, X_n)$, by taking into account also a set of independent variables $(X_{n+1}, \ldots, X_{n+m})$, representing quantities associated to the substrate and fixed by the experimenter. The general format for these equations is

$$\dot{X}_i = V_i^+(X_1, \ldots, X_{n+m}) - V_i^-(X_1, \ldots, X_{n+m}),$$

i.e. the speed of change of quantity X_i is determined by a production term (V_i^+) and a degradation term (V_i^-). Each V term contains all the variables directly influencing the behavior of X_i and it has a particular format:

$$V_i(X_1, \ldots, X_{n+m}) = \alpha_i \prod_{j=1}^{n+m} X_j^{g_{ij}},$$

where α and h_j are parameters to be determined. The general format for an S-System of differential equations is therefore:

$$\dot{X}_i = \alpha_i \prod_{j=1}^{n+m} X_j^{g_{ij}} - \beta_i \prod_{j=1}^{n+m} X_j^{h_{ij}}. \tag{3.14}$$

The parameters $\alpha_i, \beta_i \geq 0$ are called the *rate constants* and represent the basic production and degradation rates for each dependent variable. The parameters g_{ij} and h_{ij} are called kinetic orders and they represent the (kinetic) strength of the interaction of the biological entity X_j in the production or degradation of X_i. They can be positive (enhancing effect), negative (inhibition effect), or zero (no effect).

The main virtues of S-System reside in their analytical tractability (of certain specific properties, cf. [236]) and in the fact that all parameters involved have a clear biological meaning (the "kinetic strength" can be measured effectively, see [236]).

Moreover, they are extremely expressive from a dynamical point of view, i.e. they are able to encode almost all behaviours shown by ordinary differential equations. This fact can be proven analytically [236], but the argument is essentially related to the fact that the terms V^+ and V^- of S-Systems are first order approximations of general functions, near an operational point, in the logarithmic coordinate space. In fact,

$$\log(V_i(X_1, \ldots, X_{n+m})) = \log(\alpha_i) + \sum_{j=1}^{m+n} g_{ij} \log(X_j).$$

Generalized mass action equations are similar to S-Systems, with the only difference that we can have more than one V^+ or V^- term per differential equation.

Deriving S-Systems parameters from other kinetic equations. The approximation property in logarithmic coordinates suggest a simple way to deduce S-System's parameters starting from a set of other ODE modeling a biological system. All we have to do is to fix an operational point P for these equations (for instance, their steady state), and compute the kinetic order g_{ij} of a variable X_j as

$$g_{ij} = \frac{\partial \log(V^+)}{\partial X_j}\Big|_P = \frac{\partial V^+}{\partial X_j}\Big|_P \frac{X_j}{V^+}\Big|_P.$$

To compute basic production and degradation rates, if we know the production speed v_{prod} at the operational point, we can solve for α the equation $V^+(X_1^P, \ldots, X_{n+m}^P) = v_{prod}$, once we have determined all the kinetic exponents. Computation of h_{ij} and β are similar.

Example: Michaelis Menten Kinetics. Let's consider an enzymatic reaction described by the Michaelis-Menten equation of Section 3.3.1, coupled with the external feeding of substrate at a constant rate k_s and the degradation of P at rate $k_p[P]$. Therefore, the equation for this system are

$$\begin{aligned} \frac{d[P]}{dt} &= \frac{V_{max}[S]}{K+[S]} - k_p[P] \\ \frac{d[S]}{dt} &= k_s - \frac{V_{max}[S]}{K+[S]}. \end{aligned} \tag{3.15}$$

Fixing the parameters at $K = 2.5$, $V_{max} = 250$, $k_p = 0.1$ and $k_s = 1$, we have that the steady state of the system is given approximatively by $S_s = 0.01$ and $P_s = 10$. We can use this information and the mechanisms presented above to derive kinetic orders and constant rates of an S-system showing the same behavior, obtaining

$$\begin{aligned} \frac{d[P]}{dt} &= 100[S] - 0.1[P] \\ \frac{d[S]}{dt} &= 1 - 100[S]. \end{aligned} \tag{3.16}$$

In particular, the kinetic order of the substrate is approximatively equal to 1, showing a nearly linear dependence of $[P]$ on $[S]$. In Figure 3.7, we have a visual comparison of a solution of the two systems of equations, starting from the same initial condition ($[P] = [S] = 0$).

Figure 3.7: Time evolution of the system of equations (3.15) with Michaelis-Menten dynamics (**left**), and the corresponding approximating S-System (3.16) (**right**).

Example: Hill Kinetics. We can proceed like in the previous paragraph to derive an S-System describing the dynamics of the Hill's kinetics, for fixed parameters. We consider the same set of reactions as above, replacing Michaelis-Menten kinetics with Hill's one, with the Hill exponent equal to 2 (see Section 3.3.2):

$$\frac{d[P]}{dt} = \frac{V_{max}[S]^2}{K^2+[S]^2} - k_p[P]$$
$$\frac{d[S]}{dt} = k_s - \frac{V_{max}[S]^2}{K^2+[S]^2}. \tag{3.17}$$

If we set the parameters like in the previous paragraph, the approximating S-System becomes:

$$\frac{d[P]}{dt} = 39.84[S]^2 - 0.1[P]$$
$$\frac{d[S]}{dt} = 1 - 39.84[S]^2. \tag{3.18}$$

In this case, we can see how the dependence of $[P]$ on $[S]$ becomes essentially quadratic, as expected. In Figure 3.8, we have a visual comparison of the numerical integration of both systems, starting from $[P] = [S] = 0$.

Figure 3.8: Time evolution of the system of equations (3.17) with Hill dynamics with Hill coefficient 2 (**left**), and the corresponding approximating S-System (3.18) (**right**).

3.3.4 Chemical Langevin Equation

In this last section, we present a modeling approach which is somewhat in the middle between the continuous deterministic description of classical mass action kinetics and the discrete stochastic description of stochastic mass action kinetics.

If we start from the same set of chemical reactions, we can build both models and try to compare their behavior, in order to understand if and when one can forget about the stochastic ingredient (that complicates the model, also from a computational point of view), and use the simpler deterministic model to capture the essence of the biological system modeled. It turns out that the two models are generally different [102, 98], as stochastic oscillations cannot be safely dropped, especially when the population of molecules of some species has low values. In addition, we can hope that the ODEs capture the average behaviour of the stochastic system, but this is true only in few, simple cases (see again [102, 98]).

However, if we are willing to accept a continuous approximation of the discrete values representing the number of molecules, we can derive from the chemical master equation (see Section 3.1.3) a stochastic differential equation that is half way between the CTMC and the mass action ODEs. For Markov Chains describing chemical reactions, the master equation has the following form:

$$\frac{dP(x_0, t_0, x, t)}{dt} = \sum_{j=1}^{v} \left[h_j(x - S^j, c_j) P(x_0, t_0, x - S^j, t) - h_j(x, c_j) P(x_0, t_0, x, t) \right], \quad (3.19)$$

where v denotes the number of reactions of the system, S^j is a vector defining the modifications of chemical species induced by reaction j, and all other quantities are defined as in Section 3.1.3. This equation describes the time evolution of the probability distribution $P(x_0, t_0, x, t)$ on states of the system, starting from time t_0 and state x_0. This differential equation has both continuous and integer values, fact that makes it extremely difficult to solve, a part from very simple cases (this is why one has necessarily to resort to stochastic simulation methods).

From equation (3.19), we can derive a differential equation for the average value of molecule populations:

$$\frac{\partial \mathbb{E}(\mathbf{X}(t))}{\partial t} = \sum_{j=1}^{v} S^j \mathbb{E}(h_j(X_j(t), c_j)), \quad (3.20)$$

where $\mathbf{X}(t)$ is the vector of molecule species at time t. Still, this is in general not solvable, as the average $\mathbb{E}(h_j(X_j(t), c_j))$ cannot be usually computed explicitly. However, if all the reactions are monomolecular, then the h_j are linear functions, hence the equations (3.20) reduce to

$$\frac{\partial \mathbb{E}(\mathbf{X}(t))}{\partial t} = \sum_{j=1}^{v} S^j h_j(\mathbb{E}(X_j(t)), c_j), \quad (3.21)$$

which are the mass action equations of the system. If the system involves at least one bimolecular reaction, this is not true anymore.

However, under the hypothesis of having many reactions of each kind happening in short time steps, while the value of h_j remains constant (staten otherwise, if the population of chemical species are large), we can approximate chemical species with continuous variables and derive a stochastic differential equation that captures, to a certain degree of accuracy, the dynamics of the system:

$$\frac{\partial \mathbf{X}(t)}{\partial t} = \underbrace{\sum_{j=1}^{v} S^j h_j(X_j(t), c_j)}_{\alpha(\mathbf{X}(t))} + \underbrace{\sum_{j=1}^{v} S^j \sqrt{h_j(X_j(t), c_j)} \Gamma_j(t)}_{\beta(\mathbf{X}(t))\Gamma(t)}, \quad (3.22)$$

where Γ_j is a gaussian white noise term. This equation, called Chemical Langevin Equation [98, 99], is composed by two terms, a drift term $\alpha(\mathbf{X}(t))$, having the form of deterministic mass action equations, and a diffusion term $\beta(\mathbf{X}(t))\Gamma(t)$, giving the stochastic effect. If we let the number of molecules and the volume of the system go to infinity with *constant ratio*, i.e. we put ourselves in the so called *thermodynamic limit*, it can be proven that the diffusion term becomes negligible, and therefore the previous equations reduce to the deterministic mass action equations. Hence, we can say [98] that the deterministic description captures the behaviour of the system only in the thermodynamic limit (meaning for *very large populations and volumes*).

Chapter 4

Proteins and the Protein Structure Prediction Problem

Proteins are the basic building blocks of life, playing a role in almost all cellular processes. As argued in Chapter 3, their functionality depends on the set of active sites on their surface. These sites are themselves determined by the characteristic spatial configuration that a protein has. This motivates the need of understanding and predicting the shape of a protein. Unfortunately, if the experimental determination of their sequence of aminoacids is a fairly easy process, identifying a protein structure in a wet lab is a complicated, time and money consuming issue. Hence, having a way to predict in silico the native structure of a protein, given its amino acid sequence, is looked like an important breakthrough in bioinformatics. This problem, however, due to its complexity, is far away to be solved in a satisfactory way.

In this chapter we present the basic concepts regarding the structure of proteins (Section 4.1) and the physical process leading to the formation of their native shape (Section 4.2), known as *Protein Folding*.

The final part of the chapter will be devoted to the presentation of modeling simplifications (Section 4.3) and computational techniques (Section 4.4) that are used in the prediction of the structure of a protein, given its sequence of composing amino acids. This problem goes under the name of *Protein Structure Prediction*.

4.1 Proteins

A protein is a *polymer* consisting of a chain of linked monomers, called amino acids, or better amino acid residues. There are 20 different main kinds of amino acids, typically referred using three (or one) letters, as summarized in Figure 4.1. This sequence is called the *primary structure* of a protein, and it is stored in the DNA, by means of the genetic code (mapping triples of nucleotides into amino acid residues).

Each amino acid is constituted by several atoms (cf. Figure 4.1), and it is composed of two different parts. The first one, called the *backbone*, is common to all aminoacids, and it is made of a nitrogen (N) and two carbon (C) atoms. One of them is called $C\alpha$ and can be considered as the *center* of the amino acid. These three atoms are bound to two hydrogen (H) atoms and one oxygen (O) atom. While an aminoacid is not part of a chain, it has also another hydrogen atom and an OH group at the extremities of the backbone, and it loses them while linked. The other part characterizes each aminoacid, and it is

Alanine	(A)	**C**ysteine	(C)
Aspartic Acid	(D)	**Glu**tamic Acid	(E)
Phenylalanine	(F)	**G**lycine	(G)
Histidine	(H)	**I**soleucine	(I)
Lysine	(K)	**L**eucine	(L)
Methionine	(M)	**A**sparagine	(N)
Proline	(P)	**Gl**utamine	(Q)
Arginine	(R)	**S**erine	(S)
Threonine	(T)	**V**aline	(V)
Tryptophan	(W)	**T**yrosine	(Y)

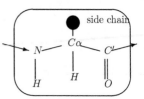

Figure 4.1: Full names of amino acids, their abbreviations, and their structure

known as *side chain*, consisting of a variable number of atoms, from 1 (Glycine) to 18 (Arginine and Tryptophan). Proteins have variable length: the smallest ones, hormones, have about 25–100 residues; typical globular proteins about 100–500; fibrous proteins may have more than 3000 residues. Thus the number of atoms involved ranges from 500 to more than 10000.

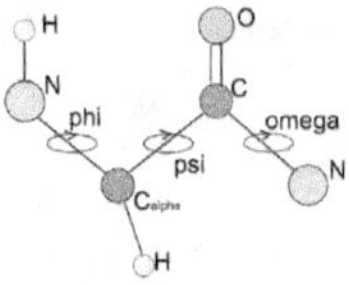

Figure 4.2: Torsional angles φ and ψ.

The links tying together the aminoacids in a protein are called *peptide bonds*, and in Figure 4.1 are represented as arrows. The first and the last aminoacid in a protein chain maintain respectively their H and OH atom. The repeating -$NC_\alpha C'$- chain of a protein is called its *backbone* (see Figure 4.4). Peptide bonds are geometrically rigid, meaning that their length is fixed, and rotations around them are not possible. On the contrary, the chain can rotate around the bonds between C_α and C' and between C_α and N, which also have a fixed length. Therefore, the *torsional angles* around these bonds, called φ and ψ, determine the entire geometry of the chain. The torsional angle ψ is defined as the angle between (the normal to) the plane spawned by $NC_\alpha C'$ and (the normal to) the plane spawned by $C_\alpha C'N$, see Figure 4.2. The torsional angle ψ is defined similarly. Also the side chains of different residues show rotational degrees of freedom, and the torsional angles involved there are denoted by indexed χ letters. Only few discrete configurations of χ angles are allowed, resulting in a finite number of possible configurations of the side chains, called *rotamers*.

A protein, once formed, is subject to interatomic forces that bend and twist the chain in a way characteristic for each protein. They cause the protein molecule to fold up into a specific three-dimensional geometric configuration called the *folded state* of the protein or *tertiary structure*, see Figure 4.3. This spatial disposition determines the chemically active groups on the surface of a folded protein, and therefore its biological function.

As said above, the spatial geometry of a protein is determined by its torsional angles φ and ψ. Plotting a ψ versus φ diagram (the so called Ramachandran plot, see Figure 4.5) for the angles extracted from known tertiary structure, we can observe that

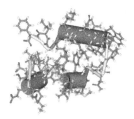

Figure 4.3: Schematic representation of the native structure of the Villin Headpiece protein, 1VII.

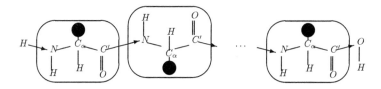

Figure 4.4: Abstract view of the amino acid backbone

only few regions are allowed. The more populated regions in that diagram correspond to local recurrent structures, known as *Secondary Structure*. There are two main classes of secondary elements: α-*helices* and β-*sheets*.

α-helices are constituted by 5 to 40 contiguous residues arranged in a regular right-handed helix with 3.6 residues per turn (left-handed helices exists, but are extremely rare), see Figure 4.7. β-sheets, instead, are constituted by extended strands of 5 to 10 residues, coupled together in parallel or antiparallel fashion, see Figure 4.6. Each strand is made of contiguous residues, but strands participating in the same sheet are not necessarily contiguous in sequence. Contiguous strands in the same sheet are usually separated by regions where the chain has an U-shaped turn, called β-*turns*. The plane containing a β-sheets can be slightly bent (β-*pleated sheets*) or closed in a cylinder (β-*barrels*). The presence of secondary structure can be predicted with high accuracy by computer programs (in [24] a program based on neural networks has a 75% accuracy, see also [208]). Secondary structure elements are usually combined together to form *motifs*, which are themselves part of *domains*, big super-secondary structures that characterize family of proteins.

Another important structural feature of proteins is the capability of cysteine residues of covalently binding through their sulphur atoms, thus forming *disulfide bridges*. This kind of information is often available, either by experiments or predictions.

The *protein structure prediction problem* is the problem of predicting the tertiary structure of a protein given its primary structure. As a protein in physiologic conditions always reaches its native state, both in vivo and in vitro, it is accepted that the primary

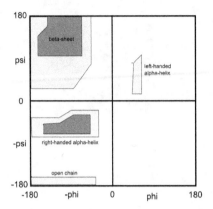

Figure 4.5: Schematic representation of the Ramachandran Plot.

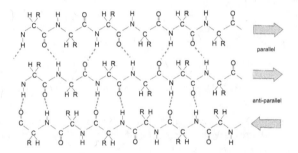

Figure 4.6: Schematic representation of parallel and anti-parallel β-strands.

structure uniquely determines the tertiary structure. Due to entropic considerations, it is also accepted that the tertiary structure is the one that minimizes the global *free energy* of the protein. This is known as *Anfisen thermodynamic hypothesis* [7]. More details are provided in the next sections.

4.2   The Physics of the Protein Folding

In this Section we give an overview of the process of folding of a protein, with particular attention on its physics. Our presentation will follow the survey of Neumaier [179], first giving motivations for the Anfisen hypothesis (Sections 4.2.1 and 4.2.2), then presenting a model for the potential energy of a protein (Section 4.2.3), and finally giving an overview of the supposed structure of the energy landscape (Section 4.2.6).

Figure 4.7: Schematic representation of an α-helix.

4.2.1   Molecular Mechanics

The motion of the atoms in a protein is obviously governed by physical laws. For every atom i, we denote its position in the three-dimensional space by the vector $\vec{x}_i = (x_{i1}, x_{i2}, x_{i3})$. The configuration of a protein, therefore, is identified by the $3N$-dimensional vector

$$\vec{x} = \begin{pmatrix} \vec{x}_1 \\ \vdots \\ \vec{x}_N \end{pmatrix} = \begin{pmatrix} x_{11} \\ x_{12} \\ x_{13} \\ \vdots \\ x_{N3} \end{pmatrix}$$

where N is the total number of atoms in the molecule. For real proteins the dimension of x is in the range of about 1500–30000.

The force balance within the molecule and the resulting dynamics can be expressed by the stochastic differential equation of the *Langevin dynamics* [143]:

$$M\ddot{x} + C\dot{x} + \nabla V(x) = D\dot{W}(t). \tag{4.1}$$

This equation describes a physical system subject to the effect of a dissipative force, proportional to velocity, and to the effect of a random force, essentially representing the effect of molecular agitation. The first term of equation (4.1) describes the change of kinetic energy, and is the product of the *mass matrix M* (diagonal) and the *acceleration* $\ddot{x}$. The second term describes the excess energy dissipated to the environment, and it is the product of a symmetric, positive definite *damping matrix* C and the velocity $\dot{x}$. The third term describes the change in the potential energy, and is expressed as the gradient of a real valued *potential function* V characteristic of the molecule. The right hand side term is a random force accounting for fluctuations due to collisions with the surrounding that dissipate the energy; it is the product of normalized white noise $\dot{W}(t)$ with a suitable matrix D. Note that the white noise depends on the temperature T of the system. More information about stochastic differential equations and Langevin dynamics can be found in [143].

The interaction with the environment is in reality much more complex, and the damping and fluctuation terms are only simplified descriptions. Improving such formulations can be cumbersome and lead quickly to mathematically intractable problems. More realism is therefore added by introducing an explicit (simplified) representation of the environment where a protein folds up, i.e. by introducing a certain number of water molecules, and extending the potential to account for their interactions with each other and with the protein's atoms. More details about the stochastic formulation of molecular mechanics can be found in [179, 166].

4.2.2   The Low Temperature Limit

In the limit for $T \to 0$, the random forces disappear, i.e. the term $D\dot{W}(t)$ vanishes. The equation (4.1) thus becomes the ordinary differential equation

$$M\ddot{x} + C\dot{x} + \nabla V(x) = 0. \tag{4.2}$$

In this case, the sum of kinetic and potential energy,

$$E = \frac{1}{2}\dot{x}^T M\dot{x} + V(x), \tag{4.3}$$

has time derivative $\dot{E} = -\dot{x}^T C\dot{x}$. Since C is positive definite, this expression is negative, and vanishes only at zero velocity, when all energy is potential energy. This means that, thanks to the dissipation term $C\dot{x}$, the molecule continually loses energy until it comes to rest at a stationary point of the potential ($\nabla V(x) = 0$). This stationary point generally is a *local minimum* of the potential.

Assuming a positive temperature, the random forces will add kinetic energy, so that the molecule will describe random oscillations around the local minimum. Sometimes, it is possible that the protein collects the necessary energy to overcome the local potential barrier and fall into another minimum, associated to a different conformation.

For rigid molecules, characterized by the fact that there is a unique minimizer in the part of state space accessible to the molecules, the minimization of the potential energy surface, produces a *stable state* which describes the geometric average shape.

Proteins, however, are not rigid, but can twist along the bonds of the backbone. Therefore, the potential energy surface is complicated and presents a very high number of local minima. It is possible that random forces allow the protein to escape from the neighborhood of one local minimum (*metastable state*) and reach another one. Such transitions are referred as *state transitions*.

The frequency of transitions depends on the temperature and on the energy barrier along the energetically most favorable path between two adjacent local minima. Any such path has its highest point on the energy surface at a peak called *transition state*. Transitions from a state with higher energy to another with lower energy are much more frequent than transitions to a higher energy level.

This implies that over long time scales, *a molecule spends most of the time in the valley close to the global minimizer of the potential.*

4.2.3   Modeling the Potential

The dynamics of atoms in a molecule is governed by the quantum theory of the participating electrons. However, for complex molecules, quantum mechanical calculations are far beyond the computational resources likely to be available in the near future. Hence chemists usually use a classical description of molecules and quantum theoretical calculations are restricted to some properties of some small parts of the molecule.

The interactions of the atoms in proteins can be classified into *bonded* and *non-bonded* interactions. The bonded interaction comprehend *Covalent bonds*, which are considered unbreakable, *Disulfide bonds*, which join together two sulfur atoms, and are slow to form and to break, and *Hydrogen bonds*, which connect two hydrogen atoms close to one oxygen atom, and are formed and broken fairly easily. Non-bonded interactions, on the other end, are the the long range *electrostatic* one, which is established between atoms carrying partial charges, and the short range *van der Waals* one, happening between all pairs of atoms.

Hydrogen bonds and non-bonded interactions are particularly relevant for the interaction of the molecule with the atoms of the solvent.

The static forces are completely specified by the potential $V(x)$. Modeling the molecule reduces to specifying the contribution of the various interactions to the potential, therefore defining the *force fields* into play. The CHARMM potential [43] is a good example of force field, composed by six kind of terms, given in the table 4.1.

$$
\begin{aligned}
V(x) = \;& \sum_{\text{bonds}} c_l(b - b_0)^2 && (b \text{ a bond length}) \\
+\;& \sum_{\text{bond angles}} c_a(\theta - \theta_0)^2 && (\theta \text{ a bond angle}) \\
+\;& \sum_{\substack{\text{improper} \\ \text{torsion angles}}} c_i(\tau - \tau_0)^2 && (\tau \text{ an improper torsion angle}) \\
+\;& \sum_{\text{dihedral angles}} \text{trig}(\omega) && (\omega \text{ a dihedral angle}) \\
+\;& \sum_{\text{charged pairs}} \frac{Q_i Q_j}{D r_{ij}} && (r_{ij} \text{ the Euclidean distance from } i \text{ to } j) \\
+\;& \sum_{\text{unbonded pairs}} c_w \varphi\left(\frac{R_i + R_j}{r_{ij}}\right) && (R_{ij} \text{ the radius of atom } i)
\end{aligned}
$$

Table 4.1: The CHARMM potential

The Q_i are *partial charges* assigned to the atoms in order to approximate the electrostatic potential of the electron cloud, and D is the dielectric constant. The quantities indexed by 0 are the reference values; different constant apply depending on the type of atom. Also the trigonometric terms $\text{trig}(\omega)$ and the *force constants* c depend on the corresponding type of atoms. Remember that a dihedral angle can be both the angle between three consecutive atoms or the torsional angle of four consecutive atoms, see Figure 4.8.

The van der Waals interactions depend on φ, which is modeled as the *Lennard-Jones*

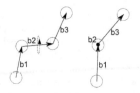

Figure 4.8: Bond and torsional dihedral angles

potential:

$$\varphi \left(\frac{R_0}{r} \right) = \left(\frac{R_0}{r} \right)^{12} - 2 \left(\frac{R_0}{r} \right)^{6}. \tag{4.4}$$

The first term decreases for small r forcing atoms to repel each other at short distance. The second term slowly increases for large r, causing an attraction of neutral atoms at large distance. There is a minimum for $r = R_0$, which models a correct behavior. Taking the equilibrium distance R_0 between two atoms as the sum $R_0 = R_i + R_j$ of the atomic radii is a simple combination rule to reduce the number of parameters.

Parameter estimation   There are several methods to determine the coefficients in a potential energy model. The principal techniques are:

- *X-ray crystallography*, giving the equilibrium positions of the atoms in crystallized proteins;

- *Nuclear magnetic resonance (NMR) spectroscopy*, giving position data of proteins in solution;

- *Ab initio* quantum mechanical calculations, giving energies, energy gradients and energy Hessians (i.e., second derivative matrices) at arbitrarily selected positions of the atoms, for molecules in the gas phase;

- Measurements of *energy spectra*, giving rather precise eigenvalues of the Hessian;

- *Thermodynamical analysis*, giving specific heats, heats of formation, conformational stability information, related to the potential in a more indirect way, via statistical mechanics.

The model parameters are adapted to data from one or several of these sources, using a mixture of least squares fitting and heuristic or interactive procedures.

4.2.4   Free Energy

The potential relevant for calculations at fixed finite temperature $T > 0$ is the *Gibbs free energy* $G = H - TS$, containing the *enthalpy* H and a correction term involving the *conformational entropy* S of the system.

The conformational entropy is proportional to the logarithm of the number of microscopically distinguishable configurations belonging to an observed macrostate and is thus roughly proportional to the logarithm of the volume of the *basin of attraction*[1] of a metastable state. Thus large flat minima (having a large catchment region) or large regions covered by many shallow local minima (corresponding to a glassy regime) are energetically more favorable than a narrow global minimum if this has only a slightly smaller potential.

Currently, entropy considerations are often addressed computationally in a qualitative fashion only, using very simple (e.g., lattice) models together with techniques from statistical mechanics, see [179].

4.2.5 The Native State

In Section 4.2.1 it is remarked that the geometry defined by the *global* minimum of the potential energy surface is expected to be the correct geometry describing the conformation observed in folded proteins. The most challenging feature of the protein folding problem is the fact that the objective function has a huge number of local minima, so that a local optimization is likely to get stuck in an arbitrary one of them, possibly far away from the desired global minimum. People working in the field expect an exponential number of local minima. Estimates range from 1.4^n to 10^n for a protein with n residues. However, most of these local minima would have a large potential energy and thus be irrelevant for global optimization; unfortunately, the number of low-lying minima seems also to grow exponentially in n.

Intuitively, the exponential number of local minima can be seen as a consequence of the combinatorial combination of locally favorable states. If we think at amino acids as single beads (cf. [179]), forgetting about the effect of non-local interactions and assuming that two consecutive beads have $m > 1$ locally stable configurations independently of other beads, we end up with a model having m^n local minima, for a chain of $n + 1$ aminoacids.

For very general energy minimization problems, combinatorial difficulty (NP-hardness) can be proved by showing that the traveling salesman problem (TSP) can be phrased as a minimization of the sum of two-body interaction energies [240], though this result does not generalize to more realistic situations.

4.2.6 The Energy Landscape

Running different experiments under physiological conditions, we can observe that most of the proteins always fold in the same native configuration. This strongly suggests the existence of a unique global minimizer with a significantly lower energy than all other local minimizers. This is also supported by experiments that suggest that the approach to the global minimum proceeds in two phases, a rapid phase to reach a nearly folded state, followed by a long period to complete the transition to the final state.

The most natural explanation is the existence of a large barrier with many (but not too many or too deep) saddles around the valley containing the global minimum. This energy landscape goes usually under the name of *folding funnel* [231].

---

[1]A basin of attraction of an equilibrium point x_0 is the region of points from which trajectories converge to x_0. It is also referred to as *catchment region*.

However, quantum mechanical corrections (accounting for vibrational energy) might disfavor the global minimum state [167, 224]. The main effect is that a slightly non-global minimum in a broader valley may be more highly populated than a global minimum in a valley with steep walls (cf. [205]). Therefore the folded state might be a metastable state with high energy barriers, or it might just be the lowest local minimizer that is kinetically accessible from most of the state space. The folded state may also correspond to more extended regions in state space where there are many close local minima of approximately the same energy as at the global minimum. This last situation corresponds to what physicists refer to as *glassy* behavior, and there are some indications from molecular dynamics simulations ([84, 132]) that this might be the situation in typical energy surfaces of proteins.

Recent studies seem to reach an agreement in that the native state is a pronounced global minimizer that is reached dynamically through a large number of transition states by an essentially *random search* through a huge set of secondary, low energy minima (representing a glassy *molten globule* state), separated from the global minimum by a large energy gap.

In simulations with simple lattice models (e.g. [77]), this scenario appears to be the necessary and sufficient conditions for folding in a reasonable time. Moreover, it also explains the so-called *Levinthal paradox* [158], stating that the time a protein needs to fold is not sufficient to explore even a tiny fraction of all local minima. However, fast collapse to a compact molten globule state, followed by a random search through the much smaller number of low energy minima, could account for the observed time scales, and the energy gap provides the stability of the native state over the molten globule state. Therefore, most of the folding time is spent trying to drive the molten globule through one of the transition states into the native geometry. On the other hand, the studies on the folding process disagree on many of the details, and the simplified drawings of the qualitative form of the potential energy surface are mutually incompatible among them.

From an evolutionary point of view, the hypothesis of a single, well separated global minimum well is also very likely. Indeed, for organisms to function successfully, the proteins performing specific tasks must fold into identical forms. Polypeptides that do not satisfy this requirement lack biological reliability and are not competitive. Thus one expects that at least the polypeptides realized as natural proteins have a single global minimum, separated from nearby local minima by a significant energy gap.

Recently, however, a number of proteins called *prions* were discovered that exist in two different folded states in nature. The normal form appears to be a metastable minimum only, separated by a huge barrier from the sick form in the global minimum. Under ordinary circumstances, only the metastable form is kinetically accessible from random states; but the presence of molecules in sick form acts as a catalyst that reduces the barrier enough to turn the normal form quickly into sick form, too. Substitution of a few crucial amino acids (caused by mutations of the prion-coding genes) also reduces the barrier.

4.3 Reduced Models of Proteins

Various simplifications of the protein folding problem are studied in the literature in order to understand the global optimization process and to simplify the development and testing of optimization algorithms.

Some of the simplified models are accurate enough so that the resulting optimal structures resemble the real native structures, and still simple enough so that the global optimization by one of the methods discussed in Chapter 5 appears feasible.

The simplifications that can be introduced fall essentially under two categories:

Detail of protein representation. One possibility is dropping information about all the atoms in the system, representing each aminoacid in a simplified way. There are different possibilities. Amino acids can be identified with *one point centered in the* C_α *atom* [232], or the *side chain* can be described as a *sphere* or an *ellipsoid* [90]. In other models, the entire backbone is represented in full atomic detail, see [223] for a detailed review.

Representation of the space of conformations. The 3D space in which amino acids are positioned can be modeled both as the *real three dimensional space* or as a *discrete lattice*. In this second case, different lattices have been used, ranging from the simple 2D and 3D cubic lattices, to Face-Centered Cubic Lattice, to lattice with an higher degree of coordination, see [108] or the paragraph hereafter for further details.

Once a representation has been chosen, a suitable energy function must be defined. This energy should be, at least in principle, capable of recognizing the native structure in the sense that this structure should ideally be at the global minimum of the energy. The choice of the representation determines also the functional form of the energy, as described hereafter.

Full energy models. Simplified full energy models have fixed bond lengths, bond angles and some torsion angles (e.g., around the peptide bond). The only degrees of freedom are an independent set of torsion angles, which limits the number of variables to $\sim 3n - 5n$, where n is the number of residues. Such models are regarded as highly reliable, but function evaluation is expensive due to the required transformations between angles and Cartesian coordinates.

Statistical backbone potential models. At the level of description of statistical potentials, only the backbone (or the backbone and a side chain center, or even only the set of C_α atoms) is modeled, with fixed bond lengths, bond angles and peptide bond torsion angles (or fixed distant of neighboring C_α's, respectively). In both cases, the number of variables is reduced to $2n$, and the potential has a simple form, determined by assuming that a set of known structures is an equilibrium ensemble of structures, so that the energy can be calculated from Boltzmann's law and statistics on the known structures. In order to obtain a useful statistics, the protein structures used must be carefully selected. The fact that the potential is now directly derived from geometric data implies that it automatically takes account of solvation and entropy corrections; on the other hand, one only gets a mean potential of less resolution. Examples of statistical potentials at this level of description are [232, 71, 159]; see [223] for a survey.

Lattice models with contact potentials. Under the choice of a discrete space, molecules are forced to have their atoms lying on lattice positions, and the potential is a sum of contact energies taken from tables derived again from statistics on databases, like [17, 175]. Now function evaluation is extremely cheap (addition of table entries for close neighbors only, resulting in a speedup factor of two order of magnitudes), and the problem has become one of combinatorial optimization. The quality of a lattice model is mainly determined by its *coordination number*, the number of permitted sites for a C_α atom in a residue adjacent to a residue with a C_α atom in a fixed position. Models used to study qualitative questions of statistical mechanics usually use a *nearest neighbor cubic lattice* (with coordination number 6). For structure prediction, a good representation of $C_\alpha - C_\alpha$ distances and angles requires at least a face-centered cubic (FCC) lattice approximation [59, 230], where α-helices and β-strands can be represented with a very low Root Mean Square Deviation (RMSD) from standard regular structures. More realistic approximations use a high coordination number of 42, 56 ([148]) or 90, with a corresponding increase in combinatorial complexity which partially offsets the gain in evaluation speed. For a review on lattice models, see [108].

4.4 Computational Approaches to Protein Structure Prediction

Protein Structure Prediction is a very fertile research area, where many computational techniques have been used. We give here a brief overview of three classes of methods, in decreasing order w.r.t. the information required other than the primary structure: homology modeling, threading and ab-initio methods.

4.4.1 Homology Modeling

Homology modeling is a technique based on the hypothesis that evolution tend to conserve both the structure of proteins and their sequence, at least in some key amino acids for the folding process. Therefore, proteins with similar sequences are supposed to have similar structures. The first step in homology modeling is to identify some target template structures. This is done by aligning the sequence of the protein versus sequences with known 3D structure in a database, identifying those with higher homology score. Then, the corresponding structures become the templates. The sequence under study is somehow fitted in the target structures, taking into account the result of the alignment. In this way, most of the aminoacids have 3D coordinates assigned. The missing ones, usually those corresponding to gaps in the alignment, are then fitted by minimizing a suitable energy function, and finally the model is completed with the remaining atoms. Homology modeling produces good results, as long as an alignment with at least 30% of homology has been found. Because of its dependence on large databases, this approach is also termed 'knowledge-based'. For a survey on homology modeling, see [103] and references therein.

4.4.2 Threading or Fold Recognition

An approach similar to homology is to try to match amino acid sequences to known folding structures. In this case, no alignment on strings is performed, but the sequence is directly aligned to the structure. This process becomes more interesting as the database of available protein geometries becomes larger and more representative. In fact, the number of different folds observed in nature is much smaller than the number of different proteins, hence there are good chances that a protein has a 3D shape close to one already known.

Various folding structures are tried in turn until one is found that makes some measure of fit (usually a statistical potential) small enough. This matching process, usually involving local optimization techniques, is called *inverse folding* or *threading*. All reasonable fitting structures are then subjected to a stability test (using molecular dynamics or Monte Carlo simulation) in order to check the correct energetic behavior of the computed structures. The decision whether a fit is reasonable must again be based on statistical potentials.

The main obstacle for successful inverse folding of general sequences resides in irregular coil regions: they can have variable length, hence the structural match must deal with insertions and deletions. Nevertheless, at present, inverse folding seems to be the most efficient way of structural prediction; and it will become more reliable as more and more proteins with known structure become available. For a survey on threading, see [170, 209].

4.4.3 Ab-Initio Methods

If the protein sequence under study has an unknown structure, both homology modeling and threading will fail in giving a reasonable model of its structure. Their limitation, in fact, resides in the fact that their accuracy depends on the current knowledge available, so new structures are out of their scope. On the contrary, ab-initio methods do not suffer from these limitations, despite having a much smaller accuracy that their knowledge-based competitors.

Ab-initio methods are all based on the Anfisen hypothesis: they search the global minimum of an energy function, for a given (reduced) representation of proteins. In principle, the most precise methods use molecular dynamics simulations [161, 199, 43]. Unfortunately, they are not feasible, due to the high intrinsic complexity of the needed operations. In particular, two to four CPU-days of a supercomputer at Berkeley Labs are needed to simulate a single nanosecond in the evolution of a medium length protein, while the typical folding time is of the order of milliseconds or seconds.

More efficient methods are offered by reduced models, using statistical potential, see Section 4.3 for further details. In this case, the conformational space needs to be searched with effective methods in order to identify the global minimizer. As commented in Section 4.2.5, this is an extremely difficult problem, given the fact that all energy functions have an exponential number of local minima to search through. In the exploration phase, different strategies can be used, ranging from local, to stochastic, to global methods, see Chapter 5 or [89] for an overview.

Among most successful ab-initio predictors (i.e. predictors which do not start from homology with a structurally characterized sequence), the programs Rosetta [40], Tasser [242], and Fragfold [136] have been performing very well in the recent CASP rounds [177]. The accuracy of the methods used in the group of Baker enabled the design (and later experimental verification) of a novel protein fold [152].

Chapter 5

Optimization Techniques

Optimization problems require to find the maximum or minimum value of a function defined over a given domain. Usually these functions take values in the real numbers, while their domains can have various nature, from natural numbers to real numbers, passing through various combinatorial structures. Formally, we have to

$$minimize\ f(\vec{x})\ with\ \vec{x} \in D,{}^{1}$$

where f is the *objective function*, D is its *feasible domain*, or *search space*, and $\vec{x} \in D$ is a *feasible point*. Usually the feasible domain D is the restriction of a bigger set induced by constraints, i.e. relations that restrict the admissible values of the variables f depends from.

Optimization problems are extremely common in applied mathematics, and there is an huge amount of research on it. The main aspect is that most of them are computationally intractable (in particular the class of combinatorial optimization problems), i.e. NP-hard (cf. [94]), leading to the development of a big variety of (approximate) algorithms to solve them.

A crucial distinction is between combinatorial optimization problems, where the domain is *discrete* and presents some "combinatorial" features, like the famous traveling salesman problem, and *continuous* optimization problems of real valued functions defined over reals. In fact, the techniques used to tackle these two classes of problems are remarkably different.

In this second case, one can hope to exploit some analytical properties of these functions in order to solve the problem. For real-valued functions defined on reals, the very basics of calculus tell us that the minimum is to be found among the points that have zero derivative or in the boundary of the domain (cf. [210]). However, these points may be difficult to identify. In every case, points of zero derivative can be local minima, for which the minimum condition is satisfied only in a small neighborhood. Global minima, on the contrary, satisfy the minimum condition over the whole domain. The identification of local and global minima are two different problems, and generally the search for local minima is much easier. Some information about techniques for the identification of local minima can be found in Section 5.1, where the famous method of gradient descent is presented.

For some classes of functions, there exist efficient computational techniques for solving the corresponding optimization problems. This is the case, for example, of linear func-

---

[1]Note that maximize $f(\vec{x})$ is the same as minimizing $-f(\vec{x})$.

tions over polyhedra on real spaces, whose minimum can be computed using the simplex algorithm (cf. [126, 219]). Convex functions can also be tackled efficiently with techniques of convex analysis (cf. always [126, 219]), but no efficient method exists for solving non-linear optimization problems in general (cf. [180] for a though overview).

For the hardest optimization problems, combinatorial ones *in primis*, many different strategies have been developed, which try to explore a small portion of the search space while finding usually solutions of good quality. Most of these algorithms, known as metaheuristics, use some stochastic ingredient, and comprehend methods like simulated annealing, taboo search and genetic algorithms. In Section 5.2 there is a description of these methodologies. More information can be found in [4, 241, 109].

In the recent years, the increasing diffusion of parallel machines on one side, and the emergence of distributed computational paradigm on the other, increased the interest in the parallel versions of these search strategies. A lot of work has been done in this direction (cf. [204, 188]), leading to parallelization of almost all search algorithms previously developed, and also to the emergence of new optimization schemes. More information can be found in Section 5.3.

As a final remark, we observe that most of these metaheuristics are designed by taking inspiration from natural phenomena. While this is very attractive, we must not forget that nature needs not to be efficient, as its inherent parallelism allows her to perform a huge number of "operations" in few instants, well beyond the power of current (and future) computers. For instance, natural processes in chemistry and biology perform a number of elementary steps in the range of 10^{15} or even more. In addition, there is no evidence that Nature "solves" optimization problems in Her daily work. Therefore, more than naturally driven methods, there is the need to identify techniques capable of exploring in an intelligent way the search space.

5.1   Local Optimization Techniques

Local Optimization techniques, also called local search algorithms (LS), are algorithms which are designed to find local minima starting from a point in their basin of attraction. There is a great number of them, and their description can be found in every introductory textbook of numerical analysis, such as [193]. For the sake of completeness, we present here probably the most basic technique, i.e. *gradient descent*.

5.1.1   Gradient Descent

Gradient descent (GD) tries to find a local minima by following the direction where the function decreases mostly. If we have to minimize a real function f, and we are in the starting point x_0, this direction is identified by $-\nabla f(x_0)$, as the gradient gives the direction of maximum growth of f.

Therefore, the updating relation is $x_{i+1} = x_i - \mu \nabla f(x_i)$, where μ is a crucial parameter influencing the convergence. If it is taken too big, the algorithm may diverge (intuitively, it jumps over the local minimum), while if taken too small, the time performance is severely damaged.

5.2 Global Optimization Methods

In this Section we present some of the most famous algorithms for solving hard optimization problems, like combinatorial ones, or minimization of complex real functions. More details can be found in [4, 202, 184, 241]. We can divide the techniques presented here in two classes: exact methods, like branch and bound, and approximate ones, going under the name of metaheuristics. In particular, we focus on these last ones, giving first a broad introduction to their class, and then focussing more specifically on some methodologies.

All metaheuristics can be seen like an *exploration of the search space*, which tries to find the global optimum using the history of their path. From this past history, and the current value of the objective function, they try to extract the information necessary to decide what to do next. Usually, there are some probabilistic ingredients at the base of these procedures. These algorithms guarantee to find a reasonable solution in a reasonable time, but in general they cannot find the global optimum.

In more detail, all these metaheuristics possess some common ingredients, i.e. they need to be fed with a definition of *neighbourhood*, with an *objective function*, with a *moving strategy* and with a *termination criterion*. In fact, all these techniques can be seen as a modification of a general search procedure called local iterative improvement (essentially a local search), which tries to improve at every step the quality of the solution found, by looking at its neighborhood. In addition they posses strategies to *escape from local minima*. All these algorithm execute by following a path in the search space, with the property that two consecutive points in this path belong one to the neighbourhood of the other. These sets are provided together with a suitable moving strategy, which explains how moves must be done in a neighborhood. Clearly, different kind of neighbourhoods and moving strategies give birth to different metaheuristics. Generally, these are designed to balance between *intensification* and *diversification*. Intensification means to search more intensely around good points, while diversification is the necessity of exploring different areas of the search space, in order not to miss good zones. Other important features in designing metaheuristics are their simplicity and easiness to develop, their robustness to the change of internal parameters and their adaptability to different problems.

Metaheuristics can be broadly divided into two classes: *exploration driven* and *population driven*. The first class contained methods where there is one point moving in the search space, like *simulated annealing* and *tabu search*. The second one, instead, keeps active a family of solutions and tries to combine them following some biologically or socially inspired criterion. The most famous representatives of this class are *genetic algorithms* (an example of *evolutionary computation*) and *ant colony optimization*.

It is worth noting that there exist packages and libraries that allow an easy development and combination of those local search procedures, like the C++ library EasyLocal++, developed by De Gaspero and Scherf [72].

5.2.1 Simulated Annealing

The idea for simulated annealing (SA) is derived from the physical process of melting and cooling a metal to form a crystalline structure. If the metal is warmed quickly and then cooled slowly (annealed), it solidify into a crystal. For physical reasons, this structure corresponds to the state of minimum free energy of the metal, so this process can be roughly seen as a kind of "minimization" performed by Nature. SA works more or less in the same way, where moves are performed using a Monte Carlo criterion. This moving strategy consists in choosing randomly a point in the neighborhood of the current position, and moving to it if the value of the objective function is improved, or with probability $\exp((f(x_{new}) - f(x_{old}))/T)$ otherwise. The parameter T is the "temperature" of the system, a control parameter governing the acceptance of hill-climbing moves, necessary not to get trapped in local minima, as they allow to jump over energy barriers. In SA, T is very high at the beginning of the simulation, and then it is gradually lowered towards zero. If the cooling is performed sufficiently slowly, i.e. following a logarithmically decreasing curve w.r.t. the number of moves, and the neighborhood satisfies some conditions,[2] then there is a theoretical guarantee that the global minimum will be found with probability one.

SA has been introduced by Kirkpatrick in [140], and a presentation from both a practical and theoretical viewpoint can be found in [3].

SA is a very common optimization heuristic also in the field of Protein Structure Prediction (see Chapter 4), where it has been used to tackle minimization problems of potential functions; see, for instance, [138, 222]

5.2.2   Genetic Algorithms

Introduced by Holland [131], genetic algorithms (GA) make use of analogies with population genetics. A whole family of "solutions" (points in the search space) is stored, and at each stage of the algorithm, this population interact and give birth to an offspring that hopefully improves the quality of the best solution. The idea is to represent each point of the search space as a collection of *chromosomes*, for instance as a sequence of bits where each bit is identified with a "chromosome". At each step of the algorithm, there is first a kind of *natural selection*, in such a way that only good solutions survive (this rule can be made more stringent in time, mimicking the effect of temperature in SA). Then the *crossover operator* is applied: it consists in exchanging part of the chromosomes between elements of the population, giving birth to new points. After that, a mutation operator is applied to each new member, to guarantee a form of "genetical variation".

The most important feature is the use of crossing over operators. This is an advantage if the used crossing rule produces offspring with a better value of the objective function (or fitness, as f is called in this community), otherwise it may be a severe disadvantage. In particular, if the different chromosomes, or coordinates, are highly correlated, the crossing over may destroy quickly good solutions, whereas if the coordinates are independent, it may be quite effective. Therefore the efficiency of GA depends dramatically on the type of crossing rule implemented, and each problem requires its own specific set of rules. In the language presented in the introduction to this section, we can say that GA have good diversification features, but poor intensification ability. That is to say, they tend to scan the whole search space, but they don't look around good candidates.

Despite these limitations, and like simulated annealing, genetic algorithm are very easy to implement, and can give quickly a decent solution to the problem. However, making them competitive requires a strong insight into the problem features. More information about genetic algorithms can be found in [12, 4, 67].

Not surprisingly, GA have been used also for PSP, see [41, 111]. An interesting algorithm for PSP is the so called *Conformational Space Annealing*, an hybrid search strategy

---

[2]Essentially, the Markov Chain underlying the system must be ergodic.

combining Simulated Annealing, Genetic Algorithms and Local Search (in order to restrict the search space to the set of local minima), see [157].

5.2.3 Tabu Search

Tabu search methods have been introduced by Glover in [104, 105], and they are a kind of neighborhood exploration, like simulated annealing. However they differ from SA mainly from the fact that they use a form of (limited) *memory* of the search, and try to exploit this information to explore better the search space. There are a lot of variants of Tabu Search (TS), but they all share some common features. At each step of the algorithm, we are in a point x of the space D, and we search its neighborhood $N(x)$ to find the best point y, i.e. the point with better value of the objective function. The only constraint is that such point must not belong to the *tabu list* T, which stores the past n points visited. If no point in $N(x)$ has a better value of x, we can move to points worsening the objective function. In this way we can escape from local minima, realizing a form of "hill-climbing". Most of TS algorithms associate to each move an *aspiration*, which is related to the goodness of the move, and which can override the tabu list constraint, improving the performances. Intensification and diversification are usually introduced here by modifying the objective function or by keeping a long term memory. Note that the tabu list avoids that the exploration path ends up in a (small) cycle. More about Tabu search methods can be found in [107, 4, 106].

5.2.4 Smoothing Methods

These optimization methodologies are suited just for minimization of real functions. They were first suggested in [225], based on the observation that in Nature the level of detail visible depends on the observation scale: if we look from far away a valley, we see a simple shape, but when we come closer, we are able to notice more and more details and roughnesses of the landscape. In fact, looking at a valley from a long distance all the small details are averaged, giving the impression of uniformity.

The hope is that doing the same for continuous functions will make us able to smooth away the details, leaving just few big valleys of local minima, making easier the task of identifying the global one. Then, gradually, the smoothing is undone, hoping that one finds itself at the global minimizer of the original surface.

While it is quite possible for such a method to miss the global minimum (so that full reliability cannot be guaranteed, and is not achieved in the tests reported by their authors), a proper implementation of this idea usually gives at least very good local minima with few function evaluations.

Smoothing Methods have been used also in PSP, see [149, 151, 150]. An interesting hybrid approach coming out this research area is the so called Quantum Mechanics Annealing, combining SA and smoothing techniques, see [6, 226, 160].

5.2.5 Exact Methods

Traditionally, e.g. [178], branch and bound (BEB) techniques are the method of choice for solving global optimization problems of a combinatorial nature, formulated as mixed integer programs. Despite all other methods, BEB algorithms are exact, in sense that they find the true global minimizer. To do this they have to explore the whole search space, or a sub-portion known to contain the global optima. The idea they exploit is to instantiate one variable at time, reducing the domains of the others according with the constraints into play, while over approximating the value of the objective function and pruning away the branches of the search tree that cannot lead to better solutions of the one found so far. If the constraint propagation and the bounding are effective, these algorithms can be quite efficient, and in addition they return the exact solution. They are also at the basis of the search engine of Constraint Logic Programming [10], which has been applied to protein structure prediction in [63], with very good results.

In addition to branch-and-bound, other exact methods include branch-and-cut and branch-and-price [178]. In addition, there are many hybrid approaches, combining, for example, integer programming, or constraint programming, with local search [198].

5.2.6   Other methods

In literature there are a lot of variants of the previously mentioned methods, and also some different approaches. Actually, the most simple one is the Multiple Random Start, which simply runs several local search algorithms starting from random chosen initial points. This algorithm can be improved by generating the starting locations using some kind of greedy rule developed accordingly to the domain into play. The obtained algorithm is called GRASP [88]. Among the variants of SA, we can mention the Threshold Accepting [81], which replaces the Monte Carlo acceptance rule with a threshold rule based on the difference in the value of the objective functions.
Other variants and metaheuristics can be found in [4, 202, 184, 241].

5.3   Parallel Optimization Methods

All the heuristics presented in Section 5.2 are instantiation of a general optimization scheme based on a local iterative improvement. The drawback is that the computational complexity of this scheme grows exponentially with the problem size [204]. A way out of this bottleneck is to try to exploit parallelism to increment the tractable dimensions. A lot of research has been done in this direction, usually in the sense of adapting some sequential search strategies to a specific parallel architecture. There is also work on general purpose parallel optimization techniques. In particular, there are some optimization algorithms which are designed as parallel from scratch, like particle swarm optimization or ant colony optimization. Good surveys are [204, 168, 187, 60, 188].

In [233] there is proposed a classification of parallel metaheuristics based on different features. In particular, they distinguish between *single walk* algorithms and *multiple walk* algorithms: in the first class there are methods where a single path on the search space is carried on, while the second class collects techniques where more than one path is followed at the same time. Among single walk algorithms there is the choice between *single step parallelism*, where neighbors are scanned simultaneously and then a move is decided, and *multiple step parallelism*, where multiple steps in the search space are made simultaneously. In the group of multiple walk parallelism, we have algorithms performing multiple *independent walks*, and others making *interacting walks*.
Finally, we must distinguish between *synchronous* and *asynchronous* parallelism: in the

first case there is a synchronization of processes governed by a common clock, while in the second there is no global time coordination.

An interesting phenomenon emerging with parallel metaheuristics is the so called criticality, discussed in [162]. It happens that the performance of a search procedure increases with the increase in the degree of parallelism up to a certain point. After that threshold, the performances of the algorithm drastically decrease to a random walk in the search space. A motivation of this fact is that the information available to every single processor can be so outdated to make them choose randomly the next move.

5.3.1 Parallel Simulated Annealing

Simulated annealing is hard to parallelize, because the converge results depend on the sequentiality of the execution. Therefore, there are two possible ways to proceed: maintain the converge by doing a "low-level" parallelization, or abandon the guarantee of asymptotic success of the algorithm. Among the first class of approaches, we can find a parallelization of the function evaluations (which is convenient only if computing the function is very costly), and a parallelization of the mechanism for generating new solutions. This is done by making several processes generate new candidate points and then choosing the best among them or the first accepted (this last technique is efficient at low temperatures).The moving mechanism of SA is kept unaltered and executed by a master processor, leaving untouched the property of convergence to a global optimum. We refer to [3, 60, 204] for a more detailed discussion.

A different approach for parallelizing SA consists in abandoning the convergence results, and introducing a kind of asynchrony in the algorithm (cf. [112] for more details). One possible direction is to let different processors follow different search paths and then merge them together after a while. Alternatively, there have been parallelization of SA where different processors updater independently different subset of variables (usually suggested by the application, as in [30] for protein structure prediction), and the information about the value of these variables is exchanged through communications of the processes. In this case errors can be introduced by the fact that the processes use outdated information for computing the value of the objective function. However, though these errors can be quantified, it is not clear under which conditions asymptotic convergence still holds, or how it is affected by the asynchrony.

5.3.2 Parallel Tabu search

Tabu search algorithm follow a path in the search space by selecting at each step the best point in the neighborhood of the current one, not forbidden by the tabu list (which stores the last n moves). There are at least two main approaches to their parallelization. The first one parallelizes the search in the neighborhood for the best solution, allowing to consider bigger neighborhoods. The second one, instead, runs several search threads, possibly starting from different initial points and possibly following different strategies. These paths can proceed independently, or can exchange information during the execution, for instance to focus the search in the most promising regions. This communication, in addition, can be performed by a regular synchronization or asynchronously. More details can be found in [60, 204].

5.3.3   Parallel Genetic Algorithms

Genetic algorithms work by applying rules of natural selection to a population of solutions (points in the search space). This means that they have to keep track of the whole population, in the process of selection and recombination. Therefore, the easiest way to parallelize them is to split the population is sub-blocks, and applying the genetic rules to these sub-populations. In addition, the quality of solutions found is incremented by exchanging members of these groups, i.e. by introducing migration between populations. This migrations can be restricted to neighboring populations. Moreover, populations can evolve asynchronously, gaining in computational time.

In [185], two different kind of parallelization of GA are distinguished: a coarse-grained one, where big sub-populations are allocated to each processor, and a fine-grained one, where each processor governs an individual or few individuals. More details about parallel genetic algorithms can be found in [60, 204].

5.3.4   Particle Swarm Optimization

Particle swarm optimization is a metaheuristic presented in [139], based on the observation of the behaviour of swarms of birds or schools of fishes, when they search for food. In fact, it happens that the communications exchanged between members of the swarms attract the whole group towards areas with more food.

The idea is to introduce a family of particles moving in the search space, and exchanging information (usually through shared memory) about the goodness of their position. These particles are characterized by two values, their position x_k and their velocity v_k. In addition, they store the best value of the objective function found in their search, f_k, plus the best global value of the objective function found so far, f_{best}.

At each execution step of the algorithm, the position and the velocity of each particle are updated using the following rules:

$$v'_k = \omega v_k + \chi_1(f_k x_k) + \chi_2(f_{best} x_k),$$

$$x'_k = x_k + v_k,$$

where χ_i are two random variables with values in $[0, 1]$ and ω is an inertia parameter, governing how much the current velocity counts in the update. In the velocity rule, the term $\chi_1(f_k x_k)$ turns the particle toward their private best (cognition part), while the term $\chi_2(f_{best} x_k)$ turns them towards the public best (social part).

The communications between particles are related to the exchange of the best values found in their search. This communication can be a broadcast to every particle, or can happen just among spatial (dynamic communication network) or relational neighbors (static communication network). More details and technicalities can be found in [139, 56].

5.3.5   Ant Colony Optimization

Ant Colony Optimization (ACO) is another nature-driven algorithm, which has been introduced in [80]. Its main idea is to use a colony of interacting agents to explore the search space. The basic mechanism of interaction is taken from the way ants cooperate, i.e. by leaving a pheromone trail in the search space. The moving strategy is probabilistic, with probabilities depending on the quantity of pheromone present in a particular point of the space. Thus the algorithm is subject to a positive feedback behaviour, as the more pheromone is in a point, the more will be added, leading to a form of autocatalysis. More information can be found in [79].

Part II

Advanced

Chapter 6

Stochastic Concurrent Constraint Programming

In this chapter we present a stochastic extension of Concurrent Constraint Programming (see Chapter 1), covering the work done in [25] and [36]. This stochastic variant of CCP will be used as a modeling language for biological systems in Chapters 7 and 8, with an eye on its relation to differential equations (Chapter 9).

The idea behind the stochastic extension is simple: we attach a stochastic duration to instructions interacting with the store, i.e. `ask` and `tell`. This duration will be represented by an exponentially distributed random variable, whose rate will depend on the context (i.e. on the current configuration of the constraint store). This feature is non-trivial, and it will exploited heavily in application to biology.

The operational semantics of the language will be given by two transition systems, the first dealing with instantaneous transitions and the second concerning stochastically timed ones. From the stochastic transition system we derive an "abstract" Markov Chain, that can be "concretized" as a discrete time Markov Chain or a continuous time one. We will prove that the discrete time chain is the jump chain of the continuous time one (see Chapter 2 for an overview on Markov Processes), and exploit this result for analyzing the long term behaviour. We will define also a notion of input output observables, both for the discrete time semantic and for the continuous one, establishing a relationship between the twos.

Finally, we present an implementation of the language, using the constraint logic programming engine on finite domains of SICStus Prolog [93].

The chapter is organized as follows: we present the syntax of the language in Section 6.1 and the operational semantics in Section 6.2. In Section 6.3 we discuss the discrete and the continuous time models and in Section 6.4 we define the notion of observables. Finally, in Section 6.5 we focus on the implementation of the language in Prolog.

6.1 Syntax

The instructions of Stochastic Concurrent Constraint Programming (sCCP) are obtained from the ones of CCP adding *rates* λ (i.e. positive real numbers) to all instructions interacting with the store, i.e. `ask` and `tell`. We can consider these rates either as a priority, inducing discrete probability distributions on the branching structure of the

$$Program = D.A$$

$$D = \varepsilon \mid D.D \mid p(\mathbf{x}) : -A$$

$$A = \mathbf{0} \mid \text{tell}_\infty(c).A \mid M \mid \exists_x A \mid A \parallel A$$
$$M = \pi.G \mid M + M$$
$$\pi = \text{tell}_\lambda(c) \mid \text{ask}_\lambda(c)$$
$$G = \mathbf{0} \mid \text{tell}_\infty(c).G \mid p(\mathbf{y}) \mid M \mid \exists_x G \mid G \parallel G$$

Table 6.1: Syntax of sCCP.

computations (through normalization), or as stochastic durations. In this second case, each instruction is associated with a continuous random variable T, representing the time needed to perform the corresponding operations in the store (i.e. adding or checking the entailment of a constraint). This random variable is exponentially distributed (cf. [181] and Section 2.2 of Chapter 2), with probability density function

$$f(\tau) = \lambda e^{-\lambda \tau},$$

where λ is a positive real number, called the rate of the exponential random variable, which can be intuitively seen as the *expected frequency per unit of time.*

The rationale behind gluing rates to `ask` and `tell` operations is, in the continuous time case, that those instructions, operating on the constraint store of the system, effectively require a certain amount of time to be executed. This time may depend on the complexity of the constraint to be told, or on the complexity of the effective computation of the entailment relation. In addition, we can think that there is a time needed by the system to compute the least upper bound operation in our constraint system, as this corresponds to a form of constraint propagation. Moreover, the time needed by these operations can depend on the current configuration of the constraint store; for instance, the more constraints are present, the harder is to compute the entailment relations, cf. [10] for a reference on constraint propagation.

In our framework, the rates associated to `ask` and `tell` are functions

$$\lambda : \mathcal{C} \to \mathbb{R}^+,$$

depending on the current configuration of the constraint store. This means that the speed of communications can vary according to the particular state of the system, though in every state of the store the random variables are perfectly defined (their rate is evaluated to a real number). Therefore, durations are sensitive to the overall status of the system and this allows to reflect locally (on communications) global properties of the model. This fact gives to the language a remarkable flexibility in modeling biological systems, see Chapter 3 for further material on this point.

The two-fold interpretation of rates either as priorities or frequency finds its motivation in the definition of a common framework where studying the relationship between discrete and continuous models of time.

The syntax of sCCP can be found in Table 6.1. An sCCP program consists in a list of procedures and in the starting configuration. Procedures are declared by specifying their name and their formal parameters, with declarations of the form $p(\mathbf{x}) : -A$. As common practice, we ask that $fv(A) \subseteq \mathbf{x}$, where $fv(A)$ denotes the free variables of agent A, as defined in Chapter 1.

Agents, on the other hand, are defined by the grammar in the last four lines of Table 6.1. There are two different actions with temporal duration, i.e. ask and tell, identified by π. Their rate λ is a function as specified above. These actions are used as guards, and can be combined together into a guarded choice M (actually, a mixed choice, as we allow both ask and tell to be combined with summation).

The A and G lines are similar, the first defining guarded agents and the second defining general agents. The only difference is that recursive call is allowed only as a guarded agent: they are instantaneous operations, thus guarding them by a timed action allows to avoid instantaneous infinite recursive loops, like those possible in $p : -A \parallel p$. In summary, an agent A can choose between different actions (M), it can perform an instantaneous tell, it can declare a variable local $(\exists_x A)$ or it can be combined sequentially ("." operator) or in parallel ($\parallel$ operator) with other agents.

Example 1. Consider the following sCCP agent

$$\exists_X \left(\text{tell}_1(X = 1) + \text{tell}_1(X = 1) \parallel \text{Bcoin}(X) \right),$$

where

$$\text{Bcoin}(X) : - \quad \text{ask}_1(X = 1).\exists_Y \left(\left(\text{tell}_{1/4}(Y = 1) + \text{tell}_{3/4}(Y = 0) \right).\text{Bcoin}(Y) \right)$$
$$+ \text{ask}_1(X = 0).\exists_Y \left(\left(\text{tell}_{1/4}(Y = 0) + \text{tell}_{3/4}(Y = 1) \right).\text{Bcoin}(Y) \right).$$

Agent $\text{Bcoin}(X)$ first checks is the X value is zero or one, and acts accordingly. Note that the main summation involves mutually exclusive ask instructions, so only one branch can be active. The internal summations update a local variable Y, setting it to 1 or 0 with different priorities, depending on the value of X. Once Y is updated, $\text{Bcoin}(X)$ calls itself recursively on variable Y. In the initial configuration of the network, the summation agent initializes the variable X, enabling the execution of agent $\text{Bcoin}(X)$.

6.1.1 Syntactical Congruence Relation

Given the grammar of Table 6.1 defining the syntax of the language, we can define the *space of syntactic processes* $\mathcal{P}_0$ as the language derivable from the grammar using either A or G as the starting variable in a derivation. $\mathcal{P}_0$ contains all possible processes that can appear during an execution of a program. In the space $\mathcal{P}_0$ we distinguish as syntactically different terms like $A \parallel B$ and $B \parallel A$. Therefore, we need to introduce axioms for the operators of the language, and this is done via a congruence relation, defined as the minimal binary relation on $\mathcal{P}_0$ closed w.r.t. rules of Table 6.2.

The first three rules simply state that the parallel operator is a commutative monoid in the space of agents, meaning that it is associative, commutative and it has agent $\mathbf{0}$ as nilpotent element. Rules four to six imply the same property for the branching operator. Rules from seven to nine deal with some basic properties of the hiding operator, namely commutativity, renaming of bound variables and removal of redundant hidings. Finally, the last rule is the absorption law for the null agent in the sequential composition.

$(CR1)$	$A_1 \parallel (A_2 \parallel A_3) \equiv (A_1 \parallel A_2) \parallel A_3$	
$(CR2)$	$A_1 \parallel A_2 \equiv A_2 \parallel A_1$	
$(CR3)$	$A_1 \parallel \mathbf{0} \equiv A_1$	
$(CR4)$	$M_1 + (M_2 + M_3) \equiv (M_1 + M_2) + M_3$	
$(CR5)$	$M_1 + M_2 \equiv M_2 + M_1$	
$(CR6)$	$M_1 + \mathbf{0} \equiv M_1$	
$(CR7)$	$\exists_x \exists_y A \equiv \exists_y \exists_x A$	
$(CR8)$	$\exists_x A \equiv \exists_y A[y/x]$	if y is not free in A
$(CR9)$	$\exists_x A \equiv A$	if x is not free in A
$(CR10)$	$A.\mathbf{0} \equiv \mathbf{0}.A \equiv A$	

Table 6.2: Congruence Relation for stochastic CCP

The congruence relation $\equiv$ is an equivalence relation partitioning the space $\mathcal{P}_0$ in classes. The set of classes $\mathcal{P} = \mathcal{P}_0/\equiv$ will be called *space of processes*. As we do not want to distinguish between syntactically congruent processes, from now on we always work with $\mathcal{P}$.

6.1.2 Configurations

Transitions of the system are determined by the agent to be executed and the current configuration of the store, thus the configuration space should be $\mathcal{P} \times \mathcal{C}$, where $\mathcal{P}$ is the space of processes defined above, and $\mathcal{C}$ is the constraint system (see Section 1.1 of Chapter 1). However, in the operational semantics we will give a rule to deal with hiding that is simpler than the usual one for CCP (see Section 1.3 of Chapter 1), cf. Section 6.2. Essentially, we replace hidden (local) variables with global variables that are fresh, i.e. never used before. For technical reasons, we need to store explicitly this set of local variables in the configuration of the system, therefore we have to add a new term in the product $\mathcal{P} \times \mathcal{C}$.

We proceed as follows. First, we split the set of variables $\mathcal{V}$ into two infinite disjoint subsets $\mathcal{V}_1$ and $\mathcal{V}_2$, $\mathcal{V}_1 \cap \mathcal{V}_2 = \emptyset$ and $\mathcal{V} = \mathcal{V}_1 \cup \mathcal{V}_2$. Then we ask that all the variables used in the definition of agents are taken from $\mathcal{V}_1$, so that all variables in $\mathcal{V}_2$ are fresh before starting a computation. Intuitively, we will use variables taken from $\mathcal{V}_2$ to replace hidden variables in an agent. In this way we can avoid name clashes with free variables or with other hidden variables. However, we need also a way to remember which variables of $\mathcal{V}_2$ we used. The simplest way is to carry this information directly in the configuration of

$(IR1)$ $\quad \langle \text{tell}_\infty(c), d, V \rangle \longrightarrow \langle \mathbf{0}, d \sqcup c, V \rangle$

$(IR2)$ $\quad \langle p(\mathbf{x}), d, V \rangle \longrightarrow \langle A[\mathbf{x}/\mathbf{y}], d, V \rangle \qquad \text{if } p(\mathbf{y}) : -A$

$(IR3)$ $\quad \langle \exists_x A, d, V \rangle \longrightarrow \langle A[y/x], d, V \cup \{y\} \rangle \quad \text{with } y \in \mathcal{V}_2 \setminus V$

$(IR4)$ $\quad \dfrac{\langle A_1, d, V \rangle \longrightarrow \langle A_1', d', V' \rangle}{\langle A_1.A_2, d, V \rangle \longrightarrow \langle A_1'.A_2, d', V' \rangle}$

$(IR5)$ $\quad \dfrac{\langle A_1, d, V \rangle \longrightarrow \langle A_1', d', V' \rangle}{\langle A_1 \parallel A_2, d, V \rangle \longrightarrow \langle A_1' \parallel A_2, d', V' \rangle}$

Table 6.3: Instantaneous transition for stochastic CCP

the system. Therefore, if we denote with $\wp_f(\mathcal{V}_2) = \mathcal{V}_f$ the collections of finite subsets of $\mathcal{V}_2$, then a configuration of the system will be a point in the space $\mathcal{P} \times \mathcal{C} \times \mathcal{V}_f$, indicated by $\langle A, d, V \rangle$. This idea of adding the set of local variables to system configurations is borrowed from [235].

It's clear that it is not really important how a local variable is called, as long as it is used just by the process that has defined it. This justifies the introduction of the following definition:

Definition 4. Let $\langle A_1, d_1, V_1 \rangle, \langle A_2, d_2, V_2 \rangle \in \mathcal{P} \times \mathcal{C} \times \mathcal{V}_f$. $\langle A_1, d_1, V_1 \rangle \equiv_{ren} \langle A_2, d_2, V_2 \rangle$ if and only if there exists a renaming $f : \mathcal{V}_2 \to \mathcal{V}_2$ such that $f(V_2) = V_1$, $d_1[f] = d_2$ and $A_2[f] \equiv A_1$.

It is simple to see that $\equiv_{ren}$ is a congruence relation (*ren* stays here for renaming). As we are not interested in the actual name given to local variables, the correct space of configurations to use is $\mathfrak{C} = (\mathcal{P} \times \mathcal{C} \times \mathcal{V}_f)/\equiv_{ren}$.

6.2 Structural Operational Semantics

The definition of the operational semantics is given specifying two different transitions: one dealing with instantaneous actions and the other with stochastically timed ones. This semantics is close to the one presented in [36], while in [25] the instantaneous transition was implicitly defined in order to deal only with local variables.

The basic idea of the operational semantics is to apply the two transitions in an interleaved way: first we apply the transitive closure of the instantaneous transition, then we do one step of the stochastic transition.

6.2.1 Instantaneous Transition

The instantaneous transition $\longrightarrow \subseteq \mathfrak{C} \times \mathfrak{C}$ is the minimal relation closed w.r.t. the rules shown in Table 6.3. Rule $(IR1)$ deals with the addition of a constraint in the store through the least upper bound operation of the lattice. Recursion is modeled by rule $(IR2)$, which substitutes the actual variables to the formal parameters in the definition of the procedure called. Rule $(IR3)$ resolves the hiding operator: each local variable x is replaced by a fresh global variable y, i.e. a variable of $\mathcal{V}_2$ not belonging to the set V of local variables already declared. Rules $(IR4)$ and $(IR5)$, finally, extend compositionally the previous rules to sequential and parallel composition. We do not need to deal with the summation at the level of instantaneous transition, because all the choices are guarded by (stochastically) timed actions.

Example 2. Consider the following sCCP agent, containing only instantaneous instructions

$$\text{tell}_\infty(X = 0).\exists_Y\,(\text{tell}_\infty(Y = 0).P(X, Y))\,,$$

where $P(Z_1, Z_2) : -A$. Using the rules of Table 6.3, we can derive the following sequence of instantaneous transitions:

$$
\begin{aligned}
&\langle \text{tell}_\infty(X = 0).\exists_Y\,(\text{tell}_\infty(Y = 0).P(X, Y))\,, true, \emptyset\rangle \longrightarrow\\
&\quad \langle \exists_Y\,(\text{tell}_\infty(Y = 0).P(X, Y))\,, (X = 0), \emptyset\rangle \longrightarrow\\
&\quad \langle \text{tell}_\infty(X_1 = 0).P(X, X_1), (X = 0), \{X_1\}\rangle \longrightarrow\\
&\quad \langle P(X, X_1), (X = 0) \sqcup (X_1 = 0), \{X_1\}\rangle \longrightarrow\\
&\quad \langle A[X, X_1/Z_1, Z_2], (X = 0) \sqcup (X_1 = 0), \{X_1\}\rangle
\end{aligned}
$$

The syntactic restrictions imposed to instantaneous actions, namely the fact that each procedure call must be under the scope of a stochastic guard, guarantee that $\longrightarrow$ can be applied only for a finite number of steps. Moreover, it is confluent.

To prove the first property, i.e. that all sequences of applications of $\longrightarrow$, starting from an arbitrary configuration, are finite, we establish the finiteness of the set of all such sequences of applications of $\longrightarrow$, called traces. This guarantees that there exists an n_0 such that all traces have length less than n_0. First of all, we need the following

Definition 5. Let $\langle A, d, V\rangle \in \mathfrak{C}$.

1. The set of *traces* $\mathcal{T}_{\langle A,d,V\rangle}$ is

$$\mathcal{T}_{\langle A,d,V\rangle} = \{\langle A, d, V\rangle \longrightarrow \langle A_1, d_1, V_1\rangle \longrightarrow \ldots \longrightarrow \langle A_n, d_n, V_n\rangle\}.$$

2. The set of *maximal traces* $\mathcal{M}_{\langle A,d,V\rangle}$ is the set

$$\mathcal{M}_{\langle A,d,V\rangle} = \{\langle A, d, V\rangle \longrightarrow \langle A_1, d_1, V_1\rangle \longrightarrow \ldots \longrightarrow \langle A_n, d_n, V_n\rangle \not\longrightarrow\}.$$

3. Let $\delta \in \mathcal{T}_{\langle A,d,V\rangle}$ be a trace. The *length* of δ, denoted by $\ell(\delta)$, is equal to n whenever $\delta = \langle A, d, V\rangle \longrightarrow \langle A_1, d_1, V_1\rangle \longrightarrow \ldots \longrightarrow \langle A_n, d_n, V_n\rangle$. In this case, the *final element* of δ is $\varphi(\delta) = \langle A_n, d_n, V_n\rangle$.

The first problem we face is whether the set $\mathcal{T}_{\langle A,d,V\rangle}$ depends on the store d or on the set of local variables V. It is easy to see that this is not the case. The only rule modifying the store is $(IR1)$, the one concerning the instantaneous addition of a constraint c. This rule gives to instantaneous tell the semantics of an *eventual tell* (see Section 1.2 of Chapter 1), i.e. the constraint is told also if it makes the store enter in an inconsistent state ($false$).

In particular, the application of this rule does not depend on the current state of the store. This is true also for all other rules of Table 6.3. Things are similar for the set of local variables: it is modified only by rule $(IR3)$, but its application does not depend on V, like that of other rules. These intuitions are formalized in the following

Lemma 1. *Let $\langle A, d, V \rangle$ and $\langle A, d', V' \rangle$ be two configurations of $\mathfrak{C}$ for the same agent A. There is a one to one correspondence between the sets $\mathcal{T}_{\langle A,d,V \rangle}$ and $\mathcal{T}_{\langle A,d',V' \rangle}$.*

Proof. Consider a trace $\delta \in \mathcal{T}_{\langle A,d,V \rangle}$, and let $n = \ell(\delta)$. We need to consider the sequence of operations of δ, i.e., for all $i \in \{1, \dots, n\}$, the rule applied in the i-th step $\langle A_{i-1}, d_{i-1}, V_{i-1} \rangle \longrightarrow \langle A_i, d_i, V_i \rangle$ (it can be one among $(IR1)$, $(IR2)$, and $(IR3)$) and the subagent of A_{i-1} it applies to. All these rules can be applied independently from the store and the set of free variables, hence they can be applied in the same order starting from $\langle A, d', V' \rangle$. In this way we generate an element of $\mathcal{T}_{\langle A,d',V' \rangle}$, thus defining a function $\Phi : \mathcal{T}_{\langle A,d,V \rangle} \to \mathcal{T}_{\langle A,d,V' \rangle}$. This function Φ is *injective*, as two different traces of $\mathcal{T}_{\langle A,d,V \rangle}$ must differ in their sequence of operations (otherwise they would be the same trace), and the same distinction propagates to the associated traces of $\mathcal{T}_{\langle A,d',V' \rangle}$.[1] Φ is also *surjective*: given a trace $\delta' \in \mathcal{T}_{\langle A,d',V' \rangle}$, we can isolate its sequence of operations and take the trace $\delta \in \mathcal{T}_{\langle A,d,V \rangle}$ with the same sequence. Clearly, δ will be mapped in δ'. ∎

Lemma 1 essentially states that set of traces depend only on the agent A of a configuration $\langle A, d, V \rangle \in \mathfrak{C}$, hence we can safely forget about d and V, writing from now on $\mathcal{T}_{\langle A \rangle}$ for $\mathcal{T}_{\langle A,d,V \rangle}$ with generic d and V.

The previous lemma allows us to prove in a simpler way the finiteness of sets $\mathcal{T}_{\langle A \rangle}$. This property is a consequence of the fact that in this language procedure calls are always guarded by stochastic actions. Therefore, after applying rule $(IR2)$, the substituted agent does not have any procedure ready to be called, but all its procedures are under a stochastic guard. Hence, at least one step of the stochastic transition must be performed in order to expose such procedure.

To prove finiteness, we proceed in two steps. Essentially, we prove first the property for a subset of the space of processes $\mathcal{P}$. With $\mathcal{P}_{ng}$ we denote the set of all the agents that are not guarded, i.e. the ones derived from the grammar of Table 6.1 using only A as starting variable.

Lemma 2. *Let $A \in \mathcal{P}_{ng}$ be a non-guarded sCCP agent. $\mathcal{T}_{\langle A \rangle}$ is finite.*

Proof. We prove the lemma by structural induction on agents $\mathcal{P}_{ng}$. We have to consider three basic cases, i.e. $A = \mathbf{0}$, $A = \text{tell}_\infty(c)$, $A = M$, and three inductive cases, i.e. $A = \exists_x A_1$, $A = A_1.A_2$, and $A = A_1 \parallel A_2$.

$A = \mathbf{0}$: Clearly $\mathcal{T}_{\langle \mathbf{0} \rangle} = \emptyset$.

$A = \text{tell}_\infty(c)$: $\mathcal{T}_{\langle \text{tell}_\infty(c) \rangle}$ contains only one element, i.e. the trace of length one obtained applying rule $(IR1)$.

---

[1] Note that, being $\mathcal{T}_{\langle A,d,V \rangle}$ defined as a set, we are not distinguishing traces with the same sequence of operations. Consider, for instance, the agent $\text{tell}_\infty(c) \parallel \text{tell}_\infty(c)$; it has two identical traces, corresponding to the different orders of execution of the first and the second agent. However, these two traces are identical, thanks to structural congruence relation, and contribute to one single element of the set $\mathcal{T}_{\langle A,d,V \rangle}$.

$A = M$: $\mathcal{T}_{\langle M \rangle}^{\longrightarrow} = \emptyset$, as no rule in Table 6.3 deals with summation.

$A = \exists_x A_1$: Traces of A are traces of A_1 prefixed by an application of rule $(IR3)$, hence the property follows by applying the inductive hypothesis on A_1.

$A = A_1.A_2$: By inductive hypothesis, $\mathcal{T}_{\langle A_1 \rangle}^{\longrightarrow}$ and $\mathcal{T}_{\langle A_2 \rangle}^{\longrightarrow}$ are both finite. Note that the only way of applying any transition to A_2 is by an application of a chain of transitions transforming A_1 to $\mathbf{0}$. Hence, the traces of $\mathcal{T}_{\langle A_1.A_2 \rangle}^{\longrightarrow}$ are those of A_1 not leading to $\mathbf{0}$ plus those leading to $\mathbf{0}$ with traces of A_2 glued to their end. The cardinality $|\mathcal{T}_{\langle A_1.A_2 \rangle}^{\longrightarrow}|$ of $\mathcal{T}_{\langle A_1.A_2 \rangle}^{\longrightarrow}$ is therefore bounded by $|\mathcal{T}_{\langle A_1 \rangle}^{\longrightarrow}| \cdot |\mathcal{T}_{\langle A_2 \rangle}^{\longrightarrow}|$.

$A = A_1 \parallel A_2$: By inductive hypothesis, $\mathcal{T}_{\langle A_1 \rangle}^{\longrightarrow}$ and $\mathcal{T}_{\langle A_2 \rangle}^{\longrightarrow}$ are both finite. A trace of $A_1 \parallel A_2$ is constructed by merging together a trace of A_1 and a trace of A_2. This merging can be done by interleaving these two traces in arbitrary manner. If the trace of A_1 has length n and the trace of A_2 has length m, the number of different interleavings is (bounded by) $\binom{m+n}{n}$. Therefore, denoting with n_0 the maximum length of a trace in $\mathcal{T}_{\langle A_1 \rangle}^{\longrightarrow}$ and with m_0 the maximum length of a trace in $\mathcal{T}_{\langle A_2 \rangle}^{\longrightarrow}$, the cardinality of $\mathcal{T}_{\langle A_1 \parallel A_2 \rangle}^{\longrightarrow}$ is bounded by $\binom{m+n}{n} \cdot |\mathcal{T}_{\langle A_1 \rangle}^{\longrightarrow}| \cdot |\mathcal{T}_{\langle A_2 \rangle}^{\longrightarrow}|$, hence finite. ∎

Lemma 3. *Let* $A \in \mathcal{P}$ *be an sCCP agent.* $\mathcal{T}_{\langle A \rangle}^{\longrightarrow}$ *is finite.*

Proof. Proof is by structural induction on A. The only additional case w.r.t. Lemma 2 is the recursive call. All the others are proven exactly in the same way. Now, let $A = p(\mathbf{y})$, where the definition of p is $p(\mathbf{y}) : -A_1$. Traces for A are obtained by first applying rule $(IR2)$, and then by gluing traces of A_1. Finiteness of $\mathcal{T}_{\langle A \rangle}^{\longrightarrow}$ follows by Lemma 2 applied to A_1. ∎

We resort now to prove confluence of transition $\longrightarrow$. To do this, we analyze the set $\mathcal{M}_{\langle A \rangle}^{\longrightarrow}$ of maximal length traces. Specifically, we prove that all traces in $\mathcal{M}_{\langle A \rangle}^{\longrightarrow}$ have the same length and terminate with the same final agent. The proof is simple, and it uses the commutative property of the least upper bound operation of the constraint system to state the immateriality of the order of adding constraints to the store by instantaneous tells. To deal with recursive call, we use the same trick of Lemma 3, first proving confluence for agents $\mathcal{P}_{ng}$.

Lemma 4. *Let* $A \in \mathcal{P}_{ng}$ *be a non-guarded sCCP agent. If* δ_1, δ_2 *are two traces of* $\mathcal{M}_{\langle A \rangle}^{\longrightarrow}$, *then* $\ell(\delta_1) = \ell(\delta_2)$ *and* $\varphi(\delta_1) = \varphi(\delta_2)$.

Proof. We prove this lemma by structural induction on A.

$A = \mathbf{0}$: Clearly $\mathcal{M}_{\langle 0 \rangle}^{\longrightarrow} = \emptyset$, hence the lemma is trivially true.

$A = \text{tell}_\infty(c)$: $\mathcal{M}_{\langle \text{tell}_\infty(c) \rangle}^{\longrightarrow}$, like $\mathcal{T}_{\langle \text{tell}_\infty(c) \rangle}^{\longrightarrow}$, contains only one element, i.e. the trace of length one obtained applying rule $(IR1)$. The lemma holds trivially also in this case.

$A = M$: $\mathcal{M}_{\langle M \rangle}^{\longrightarrow} = \emptyset$.

$A = \exists_x A_1$: Maximal traces of A are maximal traces of A_1 preceded by an application of rule $(IR3)$, hence the property follows by applying the inductive hypothesis on A_1 (names given to local variables are not important, as we are working modulo renaming).

$A = A_1.A_2$: By inductive hypothesis, the property holds for $\mathcal{M}_{\overrightarrow{\langle A_1 \rangle}}$ and $\mathcal{M}_{\overrightarrow{\langle A_2 \rangle}}$. Two cases are give: either $\varphi(\delta) = \mathbf{0}$ for $\delta \in \mathcal{M}_{\overrightarrow{\langle A_1 \rangle}}$ or $\varphi(\delta) \neq \mathbf{0}$. In the first case, maximal traces of $A_1.A_2$ are obtained by gluing a maximal trace of A_2 after a maximal trace of A_1. Note that the condition the final element of the trace implies that also the final constraint stores are equal for all maximal traces, hence by applying inductive hypothesis, we conclude that the property holds also for $A_1.A_2$. The second case, when $f(\delta) \neq \mathbf{0}$, implies that $\mathcal{M}_{\overrightarrow{\langle A_1.A_2 \rangle}} = \mathcal{M}_{\overrightarrow{\langle A_1 \rangle}}$.

$A = A_1 \parallel A_2$: By inductive hypothesis, the property holds for $\mathcal{M}_{\overrightarrow{\langle A_1 \rangle}}$ and $\mathcal{M}_{\overrightarrow{\langle A_2 \rangle}}$. A maximal trace of $A_1 \parallel A_2$ is constructed by merging together a maximal trace of A_1 and a maximal trace of A_2. This merging can be done by interleaving these two traces in arbitrary manner. In every case, traces in $\mathcal{M}_{\overrightarrow{\langle A_1 \parallel A_2 \rangle}}$ will always have the same length, i.e. $\ell(\delta_1) + \ell(\delta_2)$, with $\delta_1 \in \mathcal{M}_{\overrightarrow{\langle A_1 \rangle}}$ and $\delta_2 \in \mathcal{M}_{\overrightarrow{\langle A_2 \rangle}}$. The final configuration is also the same for all traces of $\mathcal{M}_{\overrightarrow{\langle A_1 \parallel A_2 \rangle}}$: equality of agents follows by induction, while equality of constraint stores follows by commutativity of least upper bound operation of the constraint system. Specifically, if $c_{11}, \ldots, c_{1k}$ and $c_{21}, \ldots, c_{2h}$ are the constraints told by A_1 and A_2 in a maximal trace, then $c_{11} \sqcup \ldots \sqcup c_{1k} \sqcup c_{21} \sqcup \ldots \sqcup c_{2h}$ will be the constraint added to the store by $A_1 \parallel A_2$, no matter what the order of addition of c_{ij} to the constraint store is. ∎

Lemma 5. *Let $A \in \mathcal{P}$ be an sCCP agent. If δ_1, δ_2 are two traces of $\mathcal{M}_{\overrightarrow{\langle A \rangle}}$, then $\ell(\delta_1) = \ell(\delta_2)$ and $\varphi(\delta_1) = \varphi(\delta_2)$.*

Proof. The lemma is soon proven by structural induction. All cases except recursive call are proven analogously to Lemma 4. To prove the property for recursion, we observe that a maximal trace in $\mathcal{M}_{\overrightarrow{\langle p(\mathbf{y}) \rangle}}$, with $p(\mathbf{y}) : -A$ is obtained by prefixing a maximal trace of A by an application of rule $(IR2)$. The properties follows immediately by application of Lemma 4 to the non-guarded agent A. ∎

Finally, we can combine all the previous lemmas in the following

Theorem 13. *The transition $\longrightarrow$ defined in Table 6.3 is confluent and can be applied only a finite number of steps to each configuration $\mathfrak{C}$.* ∎

To simplify the notation in the following sections, we give the following

Definition 6. *Let $\langle A, d, V \rangle \in \mathfrak{C}$ a configuration of the system. With $\overrightarrow{\langle A, d, V \rangle}$ we denote the element $\langle A', d', V' \rangle \in \mathfrak{C}$ obtained by applying the instantaneous transition to $\langle A, d, V \rangle$ until possible. Theorem 13 guaranteed that $\overrightarrow{\langle A, d, V \rangle}$ is well defined.*

$$(SR1) \qquad \langle \text{tell}_\lambda(c), d, V \rangle \Longrightarrow_{(1,\lambda(d))} \langle \mathbf{0}, d \sqcup c, V \rangle \qquad\qquad \text{if } d \sqcup c \neq false$$

$$(SR2) \qquad \langle \text{ask}_\lambda(c), d, V \rangle \Longrightarrow_{(1,\lambda(d))} \langle \mathbf{0}, d, V \rangle \qquad\qquad \text{if } d \vdash c$$

$$(SR3) \qquad \frac{\langle \pi, d, V \rangle \Longrightarrow_{(p,\lambda)} \langle \mathbf{0}, d', V \rangle}{\langle \pi.A, d, V \rangle \Longrightarrow_{(p,\lambda)} \overrightarrow{\langle A, d', V \rangle}} \qquad \begin{array}{l} \text{with } \pi = \text{ask}_\lambda(c) \\ \text{or } \pi = \text{tell}_\lambda(c) \end{array}$$

$$(SR4) \qquad \frac{\langle A_1, d, V \rangle \Longrightarrow_{(p,\lambda)} \overrightarrow{\langle A_1', d', V' \rangle}}{\langle A_1.A_2, d, V \rangle \Longrightarrow_{(p,\lambda)} \overrightarrow{\langle A_1'.A_2, d', V' \rangle}}$$

$$(SR5) \qquad \frac{\langle M_1, d, V \rangle \Longrightarrow_{(p,\lambda)} \overrightarrow{\langle A_1', d', V' \rangle}}{\langle M_1 + M_2, d, V \rangle \Longrightarrow_{(p',\lambda')} \overrightarrow{\langle A_1', d', V' \rangle}}$$
$$\text{with } p' = \frac{p\lambda}{\lambda + \text{rate}(\langle M_2, d, V \rangle)} \text{ and } \lambda' = \lambda + \text{rate}(\langle M_2, d, V \rangle)$$

$$(SR6) \qquad \frac{\langle A_1, d, V \rangle \Longrightarrow_{(p,\lambda)} \overrightarrow{\langle A_1', d', V' \rangle}}{\langle A_1 \parallel A_2, d, V \rangle \Longrightarrow_{(p',\lambda')} \overrightarrow{\langle A_1' \parallel A_2, d', V' \rangle}}$$
$$\text{with } p' = \frac{p\lambda}{\lambda + \text{rate}(\langle A_2, d, V \rangle)} \text{ and } \lambda' = \lambda + \text{rate}(\langle A_2, d, V \rangle)$$

Table 6.4: Stochastic transition relation for stochastic CCP.

6.2.2 Stochastic Transition

The stochastic transition relation $\Longrightarrow \subseteq \mathfrak{C} \times [0, 1] \times \mathbb{R}^+ \times \mathfrak{C}$ is defined as the minimal relation closed w.r.t. rules in Table 6.4. This transition is labeled by two numbers: the first one is the probability of the transition, while the second one is its global rate. Intuitively, what we are carrying in the transition relation is the characterization of a CTMC in terms of its jump chain and holding times, see Section 2.3.1 of Chapter 2 for further details. Note that dropping the information about global rates, we get a DTMC that is exactly the jump chain of the CTMC. The exact details of the generation of the underlying Markov Processes and their correspondence is postponed in Section 6.3. In this section, instead, we focus on the definition of the transition relation and on proving some basic properties of the labels attached to it.

The stochastic behaviour of the language is given by rates, which are functions associating to each configuration of the store a positive real number. This corresponds to an explicit dependence of the stochastic behaviour from the context in which agents are executed. Only two instructions have rates attached, namely the guards ask and tell. Rule $(SR1)$ deals with stochastic tell, that has here the semantic of the *atomic tell*, meaning that it can be executed only if it adding its constraint does not make the store inconsistent. Note that instantaneous tell has the semantics of the eventual tell, a necessary condition in order to guarantee confluence of the instantaneous transition. Rule $(SR1)$ works as follows: if the told constraint is consistent with the store, i.e. $c \sqcup d \neq false$, then it is added with probability 1 and global rate $\lambda(d)$, i.e. the rate of the instruction evaluated w.r.t. current configuration of the store. The behavior of ask is governed by rule $(SR2)$: the instruction is active only if the asked constraint is entailed by the current configuration of the constraint store. Whenever this happens, the computation proceeds with probability one and rate $\lambda(d)$. Rule $(SR3)$ models the behaviour of a guard: computation can proceed only if a guard is active. Rule $(SR4)$, instead, deals with sequential composition in the obvious way, i.e. by preserving unaltered the behaviour of the first instruction. Rules $(SR5)$ and $(SR6)$, finally, deal with the choice and the parallel construct. They state that, if a single term of the sum or of the parallel composition can evolve with a certain probability and a certain rate, the whole construct can evolve with a new probability and a new rate given by the expressions of Table 6.4. In the definition of these rules, we make use of the function rate, assigning to each agent its global rate and defined recursively as follows:

Definition 7. The function rate $: \mathfrak{C} \to \mathbb{R}$ is defined by

1. $\mathrm{rate}\left(\langle \mathbf{0}, d, V \rangle\right) = 0;$

2. $\mathrm{rate}\left(\langle \mathrm{tell}_\lambda(c), d, V \rangle\right) = \lambda(d);$

3. $\mathrm{rate}\left(\langle \mathrm{tell}_\lambda(c), d, V \rangle\right) = 0$ if $c \sqcup d = false;$

4. $\mathrm{rate}\left(\langle \mathrm{ask}_\lambda(c), d, V \rangle\right) = \lambda(d)$ if $d \vdash c;$

5. $\mathrm{rate}\left(\langle \mathrm{ask}_\lambda(c), d, V \rangle\right) = 0$ if $d \nvdash c;$

6. $\mathrm{rate}\left(\langle \pi.A, d, V \rangle\right) = \mathrm{rate}\left(\langle \pi, d, V \rangle\right),$ where $\pi = \mathrm{ask}$ or $\pi = \mathrm{tell};$

7. $\mathrm{rate}\left(\langle A_1.A_2, d, V \rangle\right) = \mathrm{rate}\left(\langle A_1, d, V \rangle\right);$

8. $\mathrm{rate}\left(\langle M_1 + M_2, d, V \rangle\right) = \mathrm{rate}\left(\langle M_1, d, V \rangle\right) + \mathrm{rate}\left(\langle M_2, d, V \rangle\right);$

9. $\mathrm{rate}\left(\langle A_1 \parallel A_2, d, V \rangle\right) = \mathrm{rate}\left(\langle A_1, d, V \rangle\right) + \mathrm{rate}\left(\langle A_2, d, V \rangle\right).$

This function identifies all active guards and combines their rates in order to compute the global rate of the agent. The combination mechanism, in the case of choice and parallel operators, is the sum of rates of addends or parallel agents. The same mechanism is used in rules $(SR5)$ and $(SR6)$ to compute the global rate. The probability, instead, is obtained by a normalization mechanism: first we compute the rate λ_0 of the guard resolved in the precondition of the rule, by multiplying the probability p times the global rate λ. Then we add rate(A_2, d, V) to λ and we normalize λ_0 w.r.t. this new global rate.

Example 3. Consider the following three sCCP agents:

1. $\mathrm{tell}_1(c) + \mathrm{tell}_1(d)$

2. $\mathrm{tell}_1(c) + \mathrm{tell}_1(c)$

3. $\mathrm{ask}_1(c) \parallel \mathrm{tell}_1(c)$

The following stochastic transitions can be easily derived from them:

1. $\langle \mathrm{tell}_1(c) + \mathrm{tell}_1(d), true, \emptyset \rangle \Longrightarrow_{(\frac{1}{2}, 2)} \langle \mathbf{0}, c, \emptyset \rangle$ or $\langle \mathrm{tell}_1(c) + \mathrm{tell}_1(d), true, \emptyset \rangle \Longrightarrow_{(\frac{1}{2}, 2)} \langle \mathbf{0}, d, \emptyset \rangle.$

2. $\langle \text{tell}_1(c) + \text{tell}_1(c), true, \emptyset \rangle \Longrightarrow_{(\frac{1}{2}, 2)} \langle \mathbf{0}, c, \emptyset \rangle$, which is different from the trace $\langle \text{tell}_1(c), true, \emptyset \rangle \Longrightarrow_{(1,1)} \langle \mathbf{0}, c, \emptyset \rangle$: absorption law does not work for stochastic summation.

3. $\langle \text{ask}_1(c).A \parallel \text{tell}_1(c), true, \emptyset \rangle \Longrightarrow_{(1,1)} \langle \text{ask}_1(c), c, \emptyset \rangle \Longrightarrow_{(1,1)} \langle A, c, \emptyset \rangle$.

After performing one step of the transition $\Longrightarrow$, we apply the transitive closure of $\longrightarrow$ to the resulting configuration. In this way we resolve all the instantaneous transition that are enabled after a stochastic one, guaranteeing that the only actions active after one $\Longrightarrow$ step are stochastic guards. More precisely, the following lemma holds:

Lemma 6. *After one step of the stochastic transition, the derived agent is either the null agent or a parallel composition of summations of the form $M_1.A_1 \parallel \ldots \parallel M_n.A_n$.*

Proof. Let A be the agent derived after applying the stochastic transition. We can safely assume that in A there is no redundant null agent (i.e the ones that can be removed by the congruence relation). Using congruence relation, we can also assume that A is a parallel composition of one or more sequential agents, say $A = A_1 \parallel \ldots \parallel A_n$, where no A_i is a parallel composition itself. If $n = 1$ and $A_1 = \mathbf{0}$, we are done. Otherwise, let's focus on a single agent A_i. A_i cannot be a null agent, as null agents have been removed applying the syntactical congruence. A_i cannot be neither an instantaneous tell, nor a procedure call, nor an agent of the form $\exists_x A$, as all these agents are removed with the application of the transitive closure of the instantaneous transition just after the stochastic one. Therefore, A_i can be a sequential composition or a choice. If it is a sequential composition, we can reason similarly with its first component A_{i1} and conclude that A_{i1} must be a summation. ∎

When an agent is of the kind specified in the previous lemma, we say that it is in *stochastic normal form*.

Intuitively, what happens when the stochastic relation is applied is a competition between the guards of summations that are active by the current configuration. If we call executable the agents prefixed by active guards, we can show that the rate of execution of the stochastic transition for an agent A (in stochastic normal form) is the sum over all rates of executable agents, while the probability of execution is the rate of the executed agent divided by the global rate.

Formally, we can define the multiset of executable agents for a given configuration as

Definition 8. Let $\langle A, d, V \rangle \in \mathfrak{C}$. The multiset of executable agents of $\langle A, d, V \rangle$, denoted by $\text{exec}(\langle A, d, V \rangle)$, is defined inductively by:

1. $\text{exec}(\langle \mathbf{0}, d, V \rangle) = \emptyset$;

2. $\text{exec}(\langle \text{tell}_\lambda(c), d, V \rangle) = \{\text{tell}_\lambda(c)\}$ if $c \sqcup d \neq false$;

3. $\text{exec}(\langle \text{tell}_\lambda(c), d, V \rangle) = \emptyset$ if $c \sqcup d = false$;

4. $\text{exec}(\langle \text{ask}_\lambda(c).A, d, V \rangle) = \{\text{ask}_\lambda(c)\}$ if $d \vdash c$;

5. $\mathrm{exec}(\langle \mathrm{ask}_\lambda(c).A, d, V\rangle) = \emptyset$ if $d \nvdash c$;

6. $\mathrm{exec}(\langle \pi.A, d, V\rangle) = \mathrm{exec}(\langle \pi, d, V\rangle)$;

7. $\mathrm{exec}(\langle A_1.A_2, d, V\rangle) = \mathrm{exec}(\langle A_1, d, V\rangle)$;

8. $\mathrm{exec}(\langle M_1 + M_2, d, V\rangle) = \mathrm{exec}(\langle M_1, d, V\rangle) \cup \mathrm{exec}(\langle M_2, d, V\rangle)$;

9. $\mathrm{exec}(\langle A_1 \parallel A_2, d, V\rangle) = \mathrm{exec}(\langle A_1, d, V\rangle) \cup \mathrm{exec}(\langle A_2, d, V\rangle)$;

We can extend the definition of the function rate to finite sets and multisets of configurations $S \subset \mathfrak{C}$ by putting

$$\mathrm{rate}(S) = \sum_{\langle A,d,V\rangle \in S} \mathrm{rate}(\langle A, d, V\rangle).$$

Equipped with this definition, we can prove the following:

Proposition 1. *Let $\langle A, d, V\rangle \in \mathfrak{C}$ be the current configuration, in stochastic normal form. Then the next stochastic transition executes one of the agents prefixed by a guard belonging to the set $\mathrm{exec}(\langle A, d, V\rangle)$, call it $\overline{A}$. Moreover, the probability of the transition (i.e. the first label in $\Longrightarrow$) is*

$$\frac{\mathrm{rate}(\langle \overline{A}, d, V\rangle)}{\mathrm{rate}\left(\mathrm{exec}(\langle A, d, V\rangle)\right)}, \tag{6.1}$$

and the rate associated to the transition (the second label in $\Longrightarrow$) is

$$\mathrm{rate}\left(\mathrm{exec}(\langle A, d, V\rangle)\right). \tag{6.2}$$

Proof. Observing the rules in Table 6.4, it is easy to see that the agents that can be executed are only (active) guards, exactly those collected by the function exec. Hence, the first part of the proposition is true. The relation on probability and global rates can be proven by structural induction. The base cases are the ones concerning guards. If $\langle \pi_\lambda, d, V\rangle$ is an active guard, then $E = \mathrm{exec}(\langle \pi_\lambda, d, V\rangle) = \langle \pi_\lambda, d, V\rangle$ and $\mathrm{rate}(E) = \lambda(d)$, thus the relations (6.1) and (6.2) hold. If $\langle \pi_\lambda, d, V\rangle$ is not active, then $E = \emptyset$ and $\mathrm{rate}(E) = 0$, so equations (6.1) and (6.2) are again true. The inductive case for sequential composition is trivial. Regarding parallel composition, suppose that we have $\langle A_1 \parallel A_2, d, V\rangle$, and that the equations (6.1) and (6.2) hold for $\langle A_1, d, V\rangle$ and $\langle A_2, d, V\rangle$. Suppose, in addition, that the executed instruction is one of A_1, call it π. We call $E = \mathrm{exec}(\langle A_1 \parallel A_2, d, V\rangle)$, $E_1 = \mathrm{exec}(\langle A_1, d, V\rangle)$, $E_2 = \mathrm{exec}(\langle A_2, d, V\rangle)$, $r = \mathrm{rate}(E)$, $r_1 = \mathrm{rate}(E1)$, $r_2 = \mathrm{rate}(E2)$, and $r_\pi = \mathrm{rate}(\langle \pi, d, V\rangle)$. Using the formula of rule $(SR6)$ and the induction hypothesis, the execution rate R is equal to $R = \mathrm{rate}(\langle A_1, d, V\rangle) + \mathrm{rate}(\langle A_2, d, V\rangle) = r_1 + r_2 = r$, while the execution probability p is equal to $p = \mathrm{prob}(\pi, A_1) \cdot \mathrm{rate}(\langle A_1, d, V\rangle)/\mathrm{rate}(\langle A_1, d, V\rangle) + \mathrm{rate}(\langle A_2, d, V\rangle) = (r_\pi/r_1) \cdot r_1/r_1 + r_2 = r_\pi/r$,[2] i.e. the formulas we want to prove. The case of summation is the same as the one for parallel operator. ∎

---

[2]With $\mathrm{prob}(\pi, A_1)$ we denote the probability of executing π when the current agent is A_1.

6.3 Stochastic Models

The stochastic transition relation of Table 6.4 defines a *labeled transition system*, which is a labeled graph $\mathfrak{L} = (\mathfrak{C}_s, \mathfrak{E}, \ell)$.
The nodes of $\mathfrak{L}$ are the configurations of the system $\sigma \in \mathfrak{C}_s$, in stochastic normal form, defined as

$$\mathfrak{C}_s = \{\sigma = \langle A, d, V \rangle \mid A \equiv \mathbf{0} \text{ or } A \equiv M_1.A_1 \parallel \ldots \parallel M_n.A_n\}.$$

The edges $\mathfrak{E}$ of $\mathfrak{L}$ connect two nodes σ_1 and σ_2 whenever we can derive a transition between the two configurations, i.e.

$$\mathfrak{E} = \{\langle \sigma_1, \sigma_2 \rangle \mid \sigma_1 \Longrightarrow_{p,\lambda} \sigma_2\}.$$

The labeling function $\ell : \mathfrak{E} \to [0,1] \times \mathbb{R}$ returns the label of the corresponding transition:

$$\ell(\langle \sigma_1, \sigma_2 \rangle) = (p, \lambda) \text{ if } \sigma_1 \Longrightarrow_{p,\lambda} \sigma_2.$$

We write ℓ_p for the function returning only the label related to probability and ℓ_r for the function returning only the label related to the global rate. As a matter of fact, this is a multi-graph, as we can derive more than one transition connecting two nodes. For instance, consider the case of $A = \text{tell}_\lambda(c) + \text{tell}_\lambda(c)$: here we have two transitions going from $\langle A, true, \emptyset \rangle$ to $\langle \mathbf{0}, c, \emptyset \rangle$, both with label $(\frac{1}{2}, \lambda(true))$. Therefore, $\mathfrak{E}$ is a multiset, and we refer to $\mathfrak{L}$ as a *labeled multi-transition system*.

The labeled transition system $\mathfrak{L}$ needs to be transformed in a Markov Chain in order to attach to it stochastic properties. Actually, we can associate to $\mathfrak{L}$ both a discrete time Markov chain and a continuous time Markov chain: in the first case we need to keep only the information of the labeling function ℓ_p, while in the second we need to multiply the values of ℓ_p and ℓ_r. This dichotomy in the stochastic interpretation of $\mathfrak{L}$ corresponds to the two different interpretation of rates presented in Section 6.1, either as priorities or as frequencies of execution.

The first step in constructing a probabilistic model describing the evolution of the system, be it a DTMC or a CTMC, is to collapse the multi-transition system $\mathfrak{L}$ into a *labeled simple transition system* $\mathfrak{L}' = (\mathfrak{C}_s, \mathfrak{E}', \ell')$, having at most one edge connecting two different nodes of the graph. First of all, consider two transitions exiting from a state $\sigma \in \mathfrak{C}_s$ of the graph, say $\sigma \Longrightarrow_{(p_1,\lambda_1)}$ and $\sigma \Longrightarrow_{(p_2,\lambda_2)}$. Proposition 1 guarantees that $\lambda_1 = \lambda_2$, being this the exit rate from state σ.
If $\sigma_i, \sigma_j \in \mathfrak{C}_s$ are two nodes of the multi-transition system, then we have an edge $\langle \sigma_i, \sigma_j \rangle \in \mathfrak{E}'$ whenever we have at least one edge $\langle \sigma_i, \sigma_j \rangle \in \mathfrak{E}$. The function $\ell'(\langle \sigma_i, \sigma_j \rangle) = (p, \lambda)$ is obtained in the following way: p is computed by summing all probabilities p' of transitions $\langle \sigma_i, \sigma_j \rangle \in \mathfrak{E}$, hence

$$\ell_p'(\langle \sigma_i, \sigma_j \rangle) = \sum_{\langle \sigma_i, \sigma_j \rangle \in \mathfrak{E}} \ell_p(\langle \sigma_i, \sigma_j \rangle),$$

while the exit rate λ remains the same,

$$\ell_r'(\langle \sigma_i, \sigma_j \rangle) = \ell_r(\langle \sigma_i, \sigma_j \rangle).$$

As an example, consider the simple process $A = \text{tell}_\lambda(c) + \text{tell}_\lambda(c)$, with λ the constant function $\lambda(c) = \lambda_0 \in \mathbb{R}^+$. The graph of the multi-transition system $\mathfrak{L}$ has only two nodes, $\sigma_0 = \langle A, true, \emptyset \rangle$ and $\sigma_1 = \langle \mathbf{0}, c, \emptyset \rangle$, and two edges $e_1, e_2 = \langle \sigma_0, \sigma_1 \rangle$, whose label are both

Figure 6.1: Effect of collapsing multiple edges connecting two nodes in the Labeled Transition System of $A = \text{tell}_\lambda(c) + \text{tell}_\lambda(c)$.

equal to $\ell(e_1) = \ell(e_2) = (\frac{1}{2}, \lambda_0)$. Therefore, the resulting simple transition system $\mathcal{L}'$ has only the edge $e = \langle \sigma_0, \sigma_1 \rangle$, with $\ell'(e) = (1, 2\lambda_0)$, taking into account that the apparent rate of the transition from σ_0 to σ_1 is doubled thanks to the race condition between the tell agents in the choice (see Figure 6.1).

We are now ready to define the discrete time Markov chain and the continuous time Markov chain associated to the simple labeled transition system $\mathcal{L}'$.

Definition 9. The discrete time Markov Chain $\mathcal{L}_D$ associated to the simple transition system $\mathcal{L}' = (\mathfrak{C}_s, \mathfrak{E}', \ell')$ is the DTMC with state space $\mathfrak{C}_s$ and stochastic matrix Π defined by

$$\pi_{ij} = \begin{cases} \ell'_p(\langle \sigma_i, \sigma_j \rangle) & \text{if } \langle \sigma_i, \sigma_j \rangle \in \mathfrak{E}' \\ 0 & \text{otherwise} \end{cases}$$

The matrix Π defined is indeed a stochastic matrix, as Proposition 1 guarantees that $\sum_j \pi_{ij} = 1$. We want to stress that in this setting a single agent, a tell construct say, does not have an absolute probability assigned a-priori by the user. It only has a rate representing its propensity or priority of being executed, while the actual probability of being chosen depends on the context in which it is inserted. In general, the broader this context, the smaller this probability. Note that in defining the DTMC, we are abstracting from the notion of time. If we take it into account, then we end up with the following

Definition 10. The continuous time Markov Chain $\mathcal{L}_C$ associated to the simple transition system $\mathcal{L}' = (\mathfrak{C}_s, \mathfrak{E}', \ell')$ is the CTMC with state space $\mathfrak{C}_s$ and generator matrix Q defined by

$$q_{ij} = \begin{cases} \ell'_p(\langle \sigma_i, \sigma_j \rangle) \cdot \ell'_r(\langle \sigma_i, \sigma_j \rangle) & \text{if } \langle \sigma_i, \sigma_j \rangle \in \mathfrak{E}' \\ 0 & \text{otherwise} \end{cases}$$

The discrete and the continuous time Markov Chains are tightly related, in fact $\mathcal{L}_D$ is the jump chain of $\mathcal{L}_C$, see Section 2.3.1 of Chapter 2. This is the content of the following

Theorem 14. *The discrete time Markov chain $\mathcal{L}_D$ is the jump chain of the continuous time Markov chain $\mathcal{L}_C$.*

Proof. Given a CTMC with Q-matrix Q, its jump chain is a DTMC whose stochastic matrix Π is defined by

$$\pi_{ij} = \frac{q_{ij}}{\sum_j q_{ij}}.$$

In our framework we have to distinguish two cases. First, let $e_{ij} = \langle \sigma_i, \sigma_j \rangle \in \mathfrak{E}'$, then

$$\frac{q_{ij}}{\sum_j q_{ij}} = \frac{\ell'_p(e_{ij}) \cdot \ell'_r(e_{ij})}{\sum_j \ell'_p(e_{ij}) \cdot \ell'_r(e_{ij})} = \frac{\ell'_p(e_{ij}) \cdot \ell'_r(e_{ij})}{\ell'_r(e_{ij}) \cdot \sum_j \ell'_p(e_{ij})} = \frac{\ell'_p(e_{ij})}{\sum_j \ell'_p(e_{ij})} = \ell'_p(e_{ij}) = \pi_{ij},$$

where the second equality holds because the exit rate of all transition exiting from σ_i are equal, while the fourth equality follows from the characterization of exiting probabilities of Proposition 1. The case in which $\langle \sigma_i, \sigma_j \rangle \notin \mathfrak{E}'$ is is simpler, as both q_{ij} and p_{ij} are zero. ∎

The previous theorem is useful for connecting properties of the discrete time process and the continuous time one, especially those related to the structure of the chain (like classes, recurrent and transient states, and so on) and to its asymptotic behaviour (like hitting probabilities). In fact, we know from Chapter 2 that those properties, for a CTMC, are essentially determined by its jump chain.

6.4 Observables

In this section we define a notion of input/output observables for the language. In particular, we have two different notions, one related to the discrete time semantics and the other connected with the continuous time one. The observable in the discrete setting is a probability distribution on the constraint store, that contains the long term behaviour of the program. In the continuous case, instead, the observables is a function assigning a probability distribution on the store for each instant of time. We make use of Theorem 14 to prove that the continuous time observables converge to the discrete time ones as time goes to infinity.

6.4.1 Discrete Time

First of all, let's recall that a trace of a DTMC is a sequence of states that are all connected by transitions of non-zero probability, i.e. a trace δ is an element of

$$\mathcal{T}_D = \{\sigma_{i_0} \sigma_{i_1} \dots \sigma_{i_n} \mid n \in \mathbb{N}, \pi_{i_j i_{j+1}} > 0\}.$$

The probability of a trace $\delta = \sigma_{i_0} \sigma_{i_1} \dots \sigma_{i_n}$ (conditional to its initial state being σ_{i_0}) is

$$\mathbb{P}(\delta) = \prod_{j=0}^{n-1} \pi_{i_j i_{j+1}}.$$

We say that such a trace has length n, and we denote its length by length (δ).

We can now define the I/O observables, corresponding to the input output behaviour (cf. [74]). First we have to define what is the probability of going from the state σ_I to the state σ_O, equal to $\mathbb{P}(\sigma_I \longrightarrow \sigma_O)$:

$$\mathbb{P}(\sigma_I \longrightarrow \sigma_O) = \sum \{p \mid p = \mathbb{P}(\delta), \ \delta = \sigma_I \sigma_1 \dots \sigma_n \sigma_O, \ n \geq 0\}.$$

Secondly, we get rid of the information about the sets of local variables, as we want to observe only the outcome in terms of the constraint store (we suppose that agents A and

A' are in stochastic normal form):

$$\mathbb{P}\left(\langle A,d \rangle \longrightarrow \langle A',d' \rangle \right) = \sum_{|V|=0}^{\infty} \mathbb{P}(\langle A,d,\emptyset \rangle \longrightarrow \langle A',d',V \rangle).$$

Note that we do not distinguish between V sets of the same cardinality, as they are all equivalent modulo renaming.

Definition 11. The *discrete time I/O observables* is a (sub)distribution of probability on the constraint store $\mathcal{C}$, given by:

$$\mathcal{O}_D\left(\langle A,d \rangle\right) = \{(d',p) \mid p = \mathbb{P}\left(\langle A,d \rangle \longrightarrow \langle \mathbf{0},d' \rangle\right)\}.$$

The previous distribution is in general a sub-probability, because we do not collect probability from infinite non-terminating computations and from deadlocked ones. Note that we are not considering failure computations, i.e. the ones for which the store is inconsistent (*false*). Some of these computations, in fact, may well terminate with a null agent, some may go on forever and some may enter in a deadlocked state, due to consistency check of stochastic tell. With this notion of observables we are therefore interested in capturing only the concept of terminating computation. The fact that $\mathcal{O}_D$ is a sub-probability distribution can be avoided by taking into account suspended states and using a general probabilistic constructions based on cylindric sets to extend the probability measure on infinite traces [221]. However, we do not follow that direction here, as we are satisfied by the interpretation of the loss of probability mass as the probability that the program never terminates, whatever the cause.

Example 4. We present now a simple example, in order to clarify the functioning of the observables. We consider a very simple program, composed by just one agent, i.e. $A = \text{tell}_{\lambda_1}(c) \parallel \text{ask}_{\lambda_2}(c).\text{tell}_{\lambda_3}(d) + \text{tell}_{\lambda_4}(e)$. The constraint system we have in mind here is very simple: c,d,e are the basic tokens, and all the possible combination via $\sqcup$ of them are different. As in A there isn't any hiding operator, we can safely forget the set of local variables in the configurations. When the computation starts from an empty constraint store *true*, the execution tree is depicted in Figure 6.2. Note that on the edges of the tree we have two numbers as labels: the first one is the probability associated to that transition, and the second one is the rate of the transition itself. At the beginning we can observe a race condition between two tell processes (the ask is not enabled), one with rate λ_1 and one with rate λ_4. So the probability of adding c to the store is $\frac{\lambda_1}{\lambda_1+\lambda_4}$, while e will be added with probability $\frac{\lambda_4}{\lambda_1+\lambda_4}$. In either case, the rate of transition is $\lambda_1 + \lambda_4$. Other transitions work similarly.

It is clear from the tree in Figure 6.2 that there are just two terminal states, i.e. $\langle \mathbf{0}, c \sqcup d \rangle$ and $\langle \mathbf{0}, c \sqcup e \rangle$. To compute the discrete time observables, we have to collect together all the probabilities of the traces leading to them, giving:

$$\mathcal{O}_D(\langle A,\top \rangle) = \left\{ (c \sqcup d, \frac{\lambda_1\lambda_2}{\overline{\lambda}}), (c \sqcup e, \frac{\overline{\lambda}-\lambda_1\lambda_2}{\overline{\lambda}}) \right\},$$

with $\overline{\lambda} = (\lambda_1 + \lambda_4)(\lambda_2 + \lambda_4)$. Note that this is a probability, as all the computations terminate (succesfully).

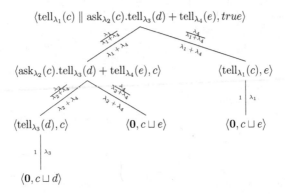

Figure 6.2: The execution tree of the program $\langle\text{tell}_{\lambda_1}(c) \parallel \text{ask}_{\lambda_2}(c).\text{tell}_{\lambda_3}(d) + \text{tell}_{\lambda_4}(e), true\rangle$

6.4.2 Continuous Time

First of all, we recall that the distribution of probability of going to a state j from a state i in time t can be reconstructed from the generator matrix Q by solving the following system of differential equations, called also *forward equation* or *stochastic master equation* of the CTMC (see Section 2.3 of Chapter 2 or [181]):

$$P'(t) = QP(t), \; P(0) = Id. \tag{6.3}$$

Given the probability matrix $P(t)$, we can give the notion of continuous time observables as a function of the elapsed time t. In fact, the probability of going from σ_I to σ_O in time t is $\mathbb{P}(\sigma_I \longrightarrow \sigma_O)(t) = P(t)_{IO}$. Abstracting from information of local variables, we have that the probability of going from a state $\langle A, d\rangle$ to a state $\langle A', d'\rangle$ in time t is

$$\mathbb{P}(\langle A, d\rangle \longrightarrow \langle A', d'\rangle)(t) = \sum_{|V|=0}^{\infty} \mathbb{P}(\langle A, d, \emptyset\rangle \longrightarrow \langle A', d', V\rangle)(t).$$

Definition 12. The *continuous time I/O observables at time t* are a sub-probability distribution over the constraint store, defined as

$$\mathcal{O}_C(\langle A, d\rangle)(t) = \{(d', p) \mid p = \mathbb{P}(\langle A, d\rangle \longrightarrow \langle \mathbf{0}, d'\rangle)(t)\}.$$

Intuitively, this gives the probability that a process terminates with final store d' in t units of time or less. This is a sub-probability because in general not all the computations have stopped after time t. We pinpoint that the notion of observable defined here is a function of the elapsed time. Actually, it corresponds to a sum of the elements of a sub-vector of the row $P(t)_{\langle A,d,\emptyset\rangle}$ of the matrix $P(t)$. An interesting question is the long term behaviour of $\mathcal{O}_C(t)$, i.e. what happens in the limit $t \to \infty$. Essentially, we are asking what is the probability of eventually reaching a terminal state. The answer is in the following

Theorem 15.

$$\lim_{t\to\infty} \mathcal{O}_C(\langle A, d\rangle)(t) = \mathcal{O}_D(\langle A, d\rangle).$$

Proof. Each state $\langle \mathbf{0}, d, V \rangle$ is an absorption state of the CTMC, meaning that when the process reaches that state, it can never leave it. The set of states $\langle \mathbf{0}, d \rangle = \{ \langle \mathbf{0}, d, V \rangle \mid |V| \in \mathbb{N} \}$ is also absorbing. The probability of ever reaching such set of states is called absorption probability, see Chapter 2. When we compute $\lim_{t \to \infty} \mathcal{O}_C(\langle A, d \rangle)(t)$, we are implicitly computing the absorption probability for each set $\langle \mathbf{0}, d \rangle$, with $d \in \mathcal{C}$. A general result about Markov Chains guarantee that this probability depends only on the jump chain of the CTMC, see Section 2.3.2 of Chapter 2. The proof is therefore concluded by applying Theorem 14. ∎

Therefore, the t-dependent observable distribution in the continuous case converges to the observable distribution defined for the discrete model of time.

This theorem connects together the long term behaviour of the discrete and the continuous time concretizations. Essentially, it states that in the asymptotic limit the additional information introduced by the continuous time model evaporates, leading to the same probability distribution induced by the discrete time model. This theorem is naturally stated and proved within this framework, which makes explicit the relations intervening between discrete and continuous interpretation of the language.

Example 5. We retake Example 4. To compute the observables in the continuous time case, we have to calculate the Q-matrix for the underlying chain. First we have to fix an ordering for the states of the program. If we set $A_3 = \langle \text{tell}_{\lambda_1}(c), e \rangle$, $A_1 = \langle \text{tell}_{\lambda_1}(c) \parallel \text{ask}_{\lambda_2}(c).\text{tell}_{\lambda_3}(d) + \text{tell}_{\lambda_4}(e), \top \rangle$, $A_4 = \langle \text{tell}_{\lambda_3}(d), c \rangle$, $A_5 = \langle \mathbf{0}, c \sqcup d \rangle$, $A_2 = \langle \text{ask}_{\lambda_2}(c).\text{tell}_{\lambda_3}(d) + \text{tell}_{\lambda_4}(e), c \rangle$, $A_6 = \langle \mathbf{0}, c \sqcup e \rangle$, with the ordering induced by the indices, we have:

$$Q = \begin{pmatrix} -\lambda_1 - \lambda_4 & \lambda_1 & \lambda_4 & 0 & 0 & 0 \\ 0 & -\lambda_2 - \lambda_4 & 0 & \lambda_2 & 0 & \lambda_4 \\ 0 & 0 & -\lambda_1 & 0 & 0 & \lambda_1 \\ 0 & 0 & 0 & -\lambda_3 & \lambda_3 & 0 \\ 0 & 0 & 0 & 0 & 0 & 0 \\ 0 & 0 & 0 & 0 & 0 & 0 \end{pmatrix}$$

Given the matrix Q, we can compute very easily the probability of going to one state to another in a certain time t, and consequently the observables. In fact, for a finite matrix, the solution of Equation (6.3) is, in terms of matrix exponentials, $P(t) = e^{tQ}$.

For instance, if $\lambda_1 = \lambda_2 = \lambda_3 = \lambda_4 = 1$, we have

$$\mathcal{O}_C(\langle A, true \rangle)(1) = \{(c \sqcup d, 0.0513), (c \sqcup e, 0.3483)\},$$

$$\mathcal{O}_C(\langle A, true \rangle)(2) = \{(c \sqcup d, 0.1467), (c \sqcup e, 0.6009)\},$$

$$\mathcal{O}_C(\langle A, true \rangle)(100) = \{(c \sqcup d, 0.2500), (c \sqcup e, 0.7500)\}.$$

As we can see, the observables converge fast to the stationary distribution given by

$$\mathcal{O}_D(\langle A, true \rangle) = \{(c \sqcup d, 0.2500), (c \sqcup e, 0.7500)\}.$$

6.5 Implementation

In the presentation of sCCP done so far we have hidden in the mechanism of the constraint store many details that cannot be ignored when one has to face the issue of implementing the language. Essentially, we need to make clear how the constraint store can be effectively managed; its definition in terms of algebraic lattice, in fact, despite being very attractive and useful for theoretical considerations, is of no use for an implementation. The constraint store has basically two operations involved: the least upper bound (addition of constraints) and the entailment relation. In addition, we need a mechanism to manage the definition of constraints that have to be added.

6.5.1 Managing the Constraint Store

Luckily, we can resort to the mechanisms developed for constraint programming, especially those used by constraint logic programming (CLP) [134]. In fact, a reasonable way to define constraints is to use prolog, exactly as in CLP, where unification and resolution are used to define complex constraints implicitly. Presenting the details of this inferential mechanism is out of the scope of this thesis, hence we refer to [134], [164] and [135], or to the example hereafter for an informal idea.

The framework of prolog is very attractive not only for the possibility of defining constraints implicitly, but also because many of its implementations have at disposal built-in libraries that manage both least upper bound and the entailment. For instance, we used the constraint logic programming library of SICStus prolog [93] on finite domains. The choice of finite domains is sufficient to use sCCP in modeling biological systems, though other domains, like reals, can be used as well. In fact, we need just to use arithmetic operations and comparisons between (stream) variables taking integer values; we can also suppose that the domain of all such variables is finite, as infinite quantities are biologically meaningless. Actually, further work is needed to identify clearly the class of constraints needed to deal with biological applications, so the definitive choice of the constraint system to be used is postponed. The CLP library of SICStus allows us to manage the constraint store: we just need to add new constraints and the built-in code takes care of updating (an approximation of) the set of possible values that variables can take. The same mechanism can also be used to compute the entailment relation: we add the negation of the constraint to be checked in the store and we control if this makes the store inconsistent. If this happens, then the entailment relation is true, otherwise is false. In every case, we have to remove the added constraint as soon as we have the answer.[3] Due to complexity reasons, the propagation mechanisms of the CLP engine are not always exact, but sometimes they compute an over-approximation of the actual domains of the variables [10]. However, we can take into account this effect by imposing that the entailment relation of the constraint store works exactly as the propagation mechanism at our disposal. This is a Byzantine way of saying that we are ignoring this problem. Note that the consistency check of the guarding tell can be implemented similarly, just by posting the constraint to be told instead of its negation.

We give now an example in order to show how the backward reasoning engine of prolog can be used to define implicitly constraints, and how this mechanism simplifies

---

[3]Essentially, we have to induce the failure of the predicate that added to the store the constraint to be checked.

considerably the writing of sCCP programs.

Suppose that in a program we could generate (through hiding operator) an arbitrary number of (*stream*) variables with domain $X \in \{1, \ldots, N\}$, and initial value $X = 1$. In addition, during the computation we want to perform a very simple operation: adding 1 to all variables defined so far except to those having value N. As these variables are stream variables, we use the special syntax defined in Section 1.6 of Chapter 1.

In sCCP, without any addition of Prolog, we have to write a family of processes, one for each number of variables that have been defined. For instance, the increase operation can be modeled as:

$$\text{increase}_n(X_1, \ldots, X_n) :\!\!-$$
$$\|_{i=1}^n \text{ask}_\lambda(X_i \neq N).\text{tell}_\infty(X_i = X_i + 1) + \text{ask}_\lambda(X_i = N).\mathbf{0}$$

Obviously, in real life one cannot write an infinite number of processes! Using Prolog, things get much simpler: we can store the variables in a list with unbounded tail[4], and then define a predicate containing this list, say `var_list(List)`. In this way, we can access to all the variables and posting new constraints using the unification of prolog, like in the predicate `increase_var` defined below.

```
increase_var :-
        var_list(List),
        max_value(N),
        increase_var(List,N).

increase_var([X,Tail],N) :-
        X<N,
        X #= X+1.
increase_var([_X,_Tail],_N).
increase_var([X|Xs],N) :-
        X<N,
        X #= X+1,
        increase_var(Xs,N).
increase_var([_X|Xs],_N) :-
        increase_var(Xs,N).
```

In the previous code, constraints are defined by the predicates X #= X+1 (we are using the syntax of CLP library of SICStus [93]), and the predicate defines a constraint that is the conjunction of the constraints X #= X+1 for all variables whose actual value is less than N. Notice that the constraint X #= X+1 is not unsatisfiable, as we are using the convention about stream variables of Section 1.6 of Chapter 1, hence its stays for a more complex predicate, whose meaning is that of adding a new element in the list associated to the stream variable X, with value equal to the last value of the list plus one. The instruction X<N does not make use of the inequality constraint #<, because it acts simply as a guard, whose failure results in the variable X not being changed. Equipped with the following predicate, the sCCP code becomes:

---

[4]alternatively to having an unbounded tail, we can use the meta-predicates assert and retract to store in the list only the variables that have been defined so far.

increase' :-

$\qquad$ tell$_{\lambda'}$(`increase_var`)

Therefore, a single sCCP predicate suffices. Note that the ask guards of the infinite family of programs have been hidden in the predicate `increase_var`.

6.5.2 The Simulation Engine

Prolog offers not only powerful built-in mechanisms to manage the store and to define constraints, but also the possibility of programming an interpreter that "executes" programs written in sCCP. In the realm of stochastic languages, executing a program means simulating the underlying Markov chain by generating one of its traces. We focussed in the implementation of the continuous time version of the language, as a discrete time trace can be obtained from a continuous time one by discarding information about time. The simulation engine is based on the Gillespie algorithm (see [102] or Section 3.1.4 of Chapter 3), which generates traces of a CTMC using a Monte Carlo mechanism exploiting their characterization in terms of Jump Chain and Holding Times. In the setting of sCCP, Gillespie algorithms works as follows. In every configuration of the system, we have several agents in parallel ready to be executed. As we suppose to have applied all possible instantaneous transitions before performing one step of the stochastic one, we can assume that the configuration of the system is in stochastic normal form, hence all the agents are (stochastic) summations. Thus, we need to test all guards of these choices, in order to see if their are active of not. In doing this, we need to evaluate and store the rate of each active guard, and then sum them up to compute the global exit rate λ from the current state. Once we did this, we choose the next agent to be executed with probability $\frac{\lambda_i}{\lambda}$, and we draw the elapsed time from an exponential distribution with rate λ. Both these operations involve the application of the simple Monte Carlo simulation method described in Chapter 2. Once we chose agent i, we need to execute it, adding the constraint to the store if it is a tell, or discarding it if it is an ask. Then the configuration of the system is updated by applying the instantaneous transition long as possible.

In order to implement effectively the previous procedure, we need to encode sCCP agents into Prolog, and lift all the operations on sCCP into operations in their Prolog description. The idea of the encoding is very simple, hence we give only a brief description. Basically, we represent all the terms of the language with their tree-like structure, and we represent a tree as a list of nested lists (essentially using a parenthesized representation of the tree). In order to work with agents modulo the syntactic congruence, we do not represent the parallel operator and the choice as a binary lists, but rather as lists of variable size. The operations of the language involving the constraint store are performed using the libraries available in Prolog, as described in the previous section. The other operations, instead, are syntactic manipulations performed at the level of the lists. For instance, to implement procedure call, we store a rule like $p(\mathbf{x})\ :\ -A$ in a predicate `rule`$(p(x), [A])$, where $[A]$ is the list representation of the agent A. The the procedure call works by replacing the list $[p(\mathbf{x})]$ with the list $[A]$. The linking of variables is performed automatically by the unification mechanism of Prolog, whenever we call the predicate `rule`$(p(x), [A])$. The selection of a branch of a choice is performed by replacing the list $[\pi_1.A_1 + \ldots + \pi_n.A_n]$ with the list of the selected branch, say $[A_i]$. We also need to merge together different lists in order to work in the space of process modulo congruence. For instance, if we execute a guard π that is followed by the agent $A = A_1 \parallel \ldots \parallel A_n$, we need to merge the list $[[A_1], \ldots, [A_n]]$ with the current list of parallel agents, after removing the list of the summation where the chosen guard belonged.

A schematic representation of the core of the interpreter is given by the predicate `stochastic_transition`.

```
stochastic_transition(Agents_List,Time) :-
    find_active_agents(Agents_List,Active_Agents_List,Global_Rate),
    choose_agent(Active_Agents_List,Global_Rate,Choosen_Agent),
    compute_elapsed_time(Time,Global_Rate,New_Time),
    execute_agent(Agents_List,Choosen_Agent,Agents_List_2),
    instantaneous_transition(Agents_List_2,New_Agents_List),
    stochastic_transition(New_Agents_List,New_Time).
```

This predicate has two input arguments: the list of agents to be executed and the current time of the simulation. First we compute the list of active agents, together with the global rate (`find_active_agents`), then we choose one of these active agents (`choose_agent`) and we update the time (`compute_elapsed_time`). After that, we update the list of agents by executing the chosen one (`execute_agent`) and by performing all possible steps of the instantaneous transition (`instantaneous_transition`). Finally, the predicate calls itself again with the new configuration.

Stream variables play an important role in the future applications of the language, hence we treat them in a special way. The classical encoding of stream variables into a logical framework like that of sCCP is as an unbounded list, whose last instantiated value represent the current value of the variable. A more efficient representation can be obtained using the meta-predicates `assert` and `retract` of Prolog: we can store the current value of a stream variable in a predicate `stream_variable(Name,Value)`, that is retracted whenever the variable needs to be updated, and then it is asserted again with the new value.

We do not pursue here the direction of proving formally the correctness of our interpreter. However, admitting that we perform correctly all the operations managing the encodings of the terms of the language, we can observe that `find_active_agents([A],L,G)` computes exactly the set $\mathrm{exec}(A)$, with $G = \mathrm{rate}(\mathrm{exec}(A))$, hence the correctness of the simulation is a straightforward consequence of Proposition 1 and the construction of the Markov chain of Section 6.3.

Chapter 7

Modeling Biological Systems in Stochastic Concurrent Constraint Programming

Taking an high level point of view, biological systems can be seen as composed essentially by two ingredients: (biological) entities and interactions among those entities. For instance, in biochemical reaction networks, the molecules are the entities and the chemical reactions are the possible interactions, see [197] and Section 3.1.1 of Chapter 3. In gene regulatory networks, instead, the entities into play are genes and regulatory proteins, while the interactions are production and degradation of proteins, and repression and enhancement of gene's expression, cf. [20] and Section 3.1.2 of Chapter 3. In addition, entities fall into two separate classes: measurable and logical. Measurable entities are those present in a certain quantity in the system, like proteins or other molecules. Logical entities, instead, have a control function (like gene gates in [20]), hence they are neither produced nor degraded. Note that logical entities are not real world entities, but rather they are part of the models: genes, in fact, are concrete objects, made of atoms. However, they are present in one single copy, hence they are intrinsically different from other molecules, fact that motivates their labeling as "logical".

The translation scheme between this nomenclature of elements and sCCP objects is summarized in Table 7.1. Measurable entities are associated to stream variables, see Section 1.6 of Chapter 1. Logical entities, instead, are represented as processes actively performing control activities. In addition, they can use variables of the constraint store either as control variables or to exchange information. Finally, each interaction is associated to a process modifying the value of certain measurable stream variables of the system.

Associating variables to measurable entities means that we are representing them as part of the environment, while the active agents are associated to the different actions capabilities of the system. These actions have a certain duration and a certain propensity to happen: a fact represented here in the standard way, i.e. associating to each action a stochastic rate. Actually, the speed of most of these actions depends on the quantity of the basic entities they act on. This fact shows clearly the need for having functional rates, which can be used to describe these dependencies explicitly.

This "interaction-centric" point of view, associating interactions to processes, is different from the usual modeling activity of stochastic process algebras, like π-calculus,

Measurable Entities	↔	Stream Variables
Logical Entities	↔	Processes (Control Variables)
Interactions	↔	Processes

Table 7.1: Schema of the mapping between elements of biological systems (left) and sCCP (right).

where each molecule is represented by a different process. The resulting approach, however, is still compositional, though what is composed together are interactions rather than molecules. In the rest of the chapter, we show that this different approach does not result in models of bigger size (we are referring to the size of the syntactic description). Actually, the flexibility inherent in sCCP allows to take also an "entity-centric" point of view: we model genes as logical entities (gene gates), hence there can be processes also associated to entities.

Concurrent constraint programming, to our knowledge, has been used to model biological systems, in its hybrid variant [22]. Hybrid CC essentially bases its semantics on hybrid automata, thus in [22] a CC-based language was used to generate hybrid models. Our approach, instead, is different, as the version of CC we use is stochastic, and so are the models considered.

Recently, some work has been done in applying a timed version of CCP [186] as a modeling language for biological systems [115]. The authors use the constraint store in a similar way as we do, though their language lacks the quantitative ingredient of rates. Moreover, they consider a discrete model of time, managed explicitly in the syntax, hence the dynamics they can describe is purely qualitative, and, in fact, their focus is in the logical analysis of possible behaviors of the system. Our approach, instead, is fundamentally quantitative, and so is the extension of CCP we propose.

Another related work worth mentioning is BIOCHAM [45], a rule-based language using constraints to define implicitly reaction rules of a biochemical system. Once the reactions are generated, different semantics can be attached to them: boolean, concentration-based (ODEs) and stochastic. Moreover, different model-checking techniques are applied to each interpretation domain. A major difference with sCCP is that our dynamics, which is always stochastic, happens at the level of the implicit syntactic description of the system given by the language, so there is no need of generating a full list of reactions (see Section 7.4 for further material).

In the following, we instantiate this general scheme to deal with two classes of biological systems: networks of biochemical reactions (Section 7.1) and genetic regulatory networks (Section 7.2). In both section, we provide some examples of systems described and analyzed in sCCP. In particular, in Section 7.1.1, we show how to use functional rates to write models where the complexity is reduced by using more complex rates (corresponding to more sophisticated chemical kinetics).

In the final part of the chapter, we try to exploit more deeply the expressivity introduced by the use of constraints and logical predicates. In particular we deal with the problem of modeling bio-regulatory networks, for which it is usually unfeasible to specify all the possible ways of reacting of all the possible reactants, due to the central role played by multimolecular complexes and protein modifications [144, 145, 147, 86, 19]. In fact, even a small set of proteins can potentially generate a big number of different complexes, of which only a few different kinds may be present at a certain time. Manually listing all these complexes can be a very hard task. We tackle this problem in a two-fold way: in Section 7.3 we show how to encode κ-calculus (see Section 3.2.4) in sCCP, while in Section 7.4 we sketch an encoding of Kohn maps (see again Section 3.2.4).

Roughly speaking, these maps work by associating a process to each complexation and modification rule. Troubles arise as those rules can be applied locally, i.e. to a sub-complex of a bigger complex, though these local changes modify the whole complex. Therefore, we need a more sophisticated way to deal with complexes: this is accomplished by exploiting some reasoning capabilities of the store, in an increasing degree of complexity from κ-calculus to Molecular Interaction Maps.

The last problem we face is that of analyzing sCCP models using stochastic model checking. As a first attempt, we use PRISM [155], an open-source stochastic model checker implementing state-of-the-art algorithms. To use PRISM, we need to define an encoding from sCCP to the language used by PRISM. Due to limitations of the latter, we need to restrict sCCP to a sub-language, with fixed number of processes in parallel during an execution (it's a limitation similar of the one introduced in PEPA, see Section 3.2.3). In Section 7.5, we present the details, providing also some examples.

7.1 Modeling Biochemical Reactions

Network of biochemical reactions are usually modeled through chemical equations of the form $R_1 + \ldots + R_n \rightarrow_k P_1 + \ldots + P_m$, where the n reactants R_i's (possibly in multiple copies) are transformed into the m products P_j's. In the equation above, either n or m can be equal to zero; the case $m = 0$ represents a degradation reaction, while the case $n = 0$ represents an external feeding of the products, performed by an experimenter. Actually, the latter is not a proper chemical reaction but rather a feature of the environmental setting, though it is convenient to represent it within the same scheme. Each reaction has an associated rate k, representing essentially its basic speed. The actual rate of the reaction is $k \cdot R_1 \cdot \ldots \cdot R_n$, where R_i denotes here the number of molecules R_i present in the system. More details on biochemical equations and the corresponding chemical kinetics can be found in Chapter 3.

There are cases when a more complex expression for the rate of the reaction is needed, see [220, 201] for further details. For instance, one may wish to describe an enzymatic reaction using a Michaelis-Menten kinetic law [83], rather than modeling explicitly the enzyme-substrate complex formation (as simple interaction/communication among molecules, cf. example below). This approximation can be done not only in deterministic modeling, but also in the context of stochastic simulation, see again [201] and Section 7.1.1.

A set of different *biochemical arrows* (corresponding to different biochemical laws) is shown in Table 7.2; this list is not exhaustive, but rather a subset of the one presented in [220]. However, adding further arrows is almost always straightforward.

In Table 7.2, we also show how to translate biochemical reactions into sCCP processes.

$$R_1 + \ldots + R_n \rightarrow_k P_1 + \ldots + P_m$$

$$\text{reaction}(k, [R_1, \ldots, R_n], [P_1, \ldots, P_m]) : -$$
$$\text{ask}_{r_{MA}(k,R_1,\ldots,R_n)} \left(\bigwedge_{i=1}^{n} (R_i > 0) \right).$$
$$\left(\|_{i=i}^{n} \text{tell}_\infty(R_i = R_i - 1) \| \right.$$
$$\left. \|_{j=1}^{m} \text{tell}_\infty(P_j = P_j + 1) \right).$$
$$\text{reaction}(k, [R_1, \ldots, R_n], [P_1, \ldots, P_m])$$

$$R_1 + \ldots + R_n \rightleftharpoons_{k_2}^{k_1} P_1 + \ldots + P_m$$

$$\text{reaction}(k_1, [R_1, \ldots, R_n], [P_1, \ldots, P_m]) \|$$
$$\text{reaction}(k_2, [P_1, \ldots, P_m], [R_1, \ldots, R_n])$$

$$S \mapsto_{K,V_0}^{E} P$$

$$\text{mm\_reaction}(K, V_0, S, P) : -$$
$$\text{ask}_{r_{MM}(K,V_0,S)}(S > 0).$$
$$(\text{tell}_\infty(S = S - 1) \| \text{tell}_\infty(P = P + 1)).$$
$$\text{mm\_reaction}(K, V_0, S, P)$$

$$S \mapsto_{K,V_0,h}^{E} P$$

$$\text{hill\_reaction}(K, V_0, h, S, P) : -$$
$$\text{ask}_{r_{Hill}(K,V_0,h,S)}(S > 0).$$
$$(\text{tell}_\infty(S = S - h) \| \text{tell}_\infty(P = P + h)).$$
$$\text{Hill\_reaction}(K, V_0, h, S, P)$$

where
$$r_{MA}(k, X_1, \ldots, X_n) = k \cdot X_1 \cdots X_n;$$
$$r_{MM}(K, V_0, S) = \frac{V_0 S}{S + K};$$
$$r_{Hill}(k, V_0, h, S) = \frac{V_0 S^h}{S^h + K^h}.$$

Table 7.2: Translation into sCCP of different biochemical reaction types, taken from the list of [220]. The reaction process models a mass-action-like reaction. It takes in input the basic rate of the reaction, the list of reactants, and the list of products. These list can be empty, corresponding to degradation and external feeding. The process has a blocking guard that checks if all the reactants are present in the system. The rate of the ask is exactly the global rate of the reaction. If the process overcomes the guard, it modifies the quantity of reactants and products and then it calls itself recursively. The reversible reaction is modeled as the combination of binding and unbinding. The third arrow corresponds to a reaction with Michaelis-Menten kinetics. The corresponding process works similarly to the first reaction, but the rate function is different. Here, in fact, the rate function is the one expressing Michaelis-Menten kinetics. See Section 7.1.1 for further details. The last arrow replaces Michaelis-Menten kinetics with Hill's one (see end of Section 7.1.1).

The basic reaction $R_1 + \ldots + R_n \rightarrow_k P_1 + \ldots P_m$ is associated to a process that first checks if all the reactants needed are present in the system (asking if all R_i are greater than zero), then it modifies the variables associated to reactants and products, and finally it calls itself recursively. Note that all the **tell** instructions have infinite rate, hence they are instantaneous transitions. The rate governing the speed of the reaction is the one associated to **ask** instruction. This rate is nothing but the function $r_{MA}(k, X_1, \ldots, X_n) = k \cdot X_1 \cdots X_n$

$$\text{enz\_reaction}(k_1, k_{-1}, k_2, S, E, ES, P) :\text{-}$$
$$\text{reaction}(k_1, [S, E], [ES]) \parallel \text{reaction}(k_{-1}, [ES], [E, S]) \parallel \text{reaction}(k_2, [ES], [E, P]).$$

$$\text{enz\_reaction}(k_1, k_{-1}, k_2, S, E, ES, P) \parallel \text{reaction}(k_{prod}, [], [S]) \parallel \text{reaction}(k_{deg}, [P], [])$$

Table 7.3: sCCP program for an enzymatic reaction with mass action kinetics. The first block defines the predicate enz_reaction($k_1, k_{-1}, k_2, S, E, ES, P$), while the second block is the definition of the entire program. The predicate reaction has been defined in Table 7.2.

representing mass action dynamics. Note that $\rightleftharpoons$ is a shorthand for the forward and the backward reactions. The arrow $\mapsto_{K,V_0}^{E}$ has a different dynamics, namely Michaelis-Menten kinetics: $r_{MM}(K, V_0, S) = \frac{V_0 S}{S+K}$. This reaction approximates the conversion of a substrate into a product due to the catalytic action of enzyme E when the substrate is much more abundant than the enzyme (quasi-steady state assumption, cf. [201, 83]). The last arrow, instead, is associated to Hill's kinetics. The dynamics represented here is an improvement on the Michaelis-Menten law, where the exponent h encodes some information about cooperativity of enzymatic binding.

Comparing the encoding of biochemical reaction into sCCP with the encoding into other process algebras like π-calculus [197], we note that the presence of functional rates gives more flexibility in the modeling phase. In fact, this kind of rates allows to describe dynamics that are different from mass action. Notable examples are Michaelis-Menten and Hill kinetics, represented by the last two arrows of Table 7.2. Notice that it is not possible to describe such kinetics in SPA where only constant rates are present, as the definition of the operational semantics constrain the dynamics to be Mass-Action like. More comments about this fact can be found in Chapter 9.

7.1.1   Example: Enzymatic Reaction

As a first and simple example, we show the model of an enzymatic reaction. We provide two different descriptions, one using a mass action kinetics, the other using a Michaelis-Menten one, see Table 7.2.

In the first case, we have the following set of reactions:

$$S + E \rightleftharpoons_{k_{-1}}^{k_1} ES \rightarrow_{k_2} P + E; \quad P \rightarrow_{k_{deg}}; \quad \rightarrow_{k_{prod}} S, \quad (7.1)$$

corresponding to a description of an enzymatic reaction that takes into account also the enzyme-substrate complex formation. Specifically, substrate S and enzyme E can bind and form the complex ES. This complex can either dissociate back into E and S, or be converted into the product P and again enzyme E. Moreover, in this particular system we added degradation of P and external feeding of S, in order to have continuous production of P. The sCCP model of this reaction can be found in Table 7.3. It is simply composed by 5 reaction agents, one for each arrow of the equations (7.1). The three reactions involving the enzyme are grouped together under the predicate enz_reaction(k1,k-1,k2,S,E,ES,P), that will be used in following subsections.

Simulations were performed with the simulator described in Chapter 6, and the trend of product P is plotted in Figure 7.1 (top). Parameters of the system were chosen in order to have, at regime, almost all the enzyme molecules in the complexed state, see caption of Figure 7.1 (top) for details.

For this simple enzymatic reaction, the quasi-steady state assumption [83] can be applied, thanks to the fact that the number of molecules of the substrate is much bigger than the number of molecules of the enzyme (meaning that almost all molecules of the enzyme are in the complexed form). Therefore, replacing the substrate-enzyme complex formation with a Michaelis-Menten kinetics (see Section 3.3.1 of Chapter 3) should leave the system behaviour unaltered. This intuition is confirmed by Figure 7.1 (bottom), showing the plot of the evolution over time of product P for the following system of reactions:

$$S \mapsto_{K,V_0}^{E} P; \qquad P \to_{k_{deg}}; \qquad \to_{k_{prod}} S,$$

whose sCCP code can be derived easily from Table 7.2.

Actually, the two graphs in Figure 7.1 are similar not just by chance, but rather as a consequence of a general result of Rao [201] about the use of steady state assumption in stochastic simulation. This can be seen as an approximate technique for model simplification, that reduces the dimension of the state space by averaging out the behaviour of some variables of the system. This averaging procedure induces a complication of the form of the stochastic rates of the models. In the case of Michaelis-Menten dynamics, it can be shown [201] that the stochastic and the deterministic form for the rates under quasi-steady state assumption coincide; this gives a formal justification for the adoption of the simplified model shown above. In [201], the authors claim also that this correspondence in the form of rates should hold also for other, more complicated, approximations induced by quasi-steady state assumption, like that of Hill's equations.

Hill's equation treat the case in which some level of cooperativity of the enzyme is to be modeled. The set of reactions in this case is an extension of the above one and, for a cooperative effect of degree two, can be written as:

$$\begin{aligned}
S + E &\rightleftharpoons_{k_{-1}}^{k_1} C_1 \to_{k_2} P + E; \\
S + C_1 &\rightleftharpoons_{k_{-3}}^{k_3} C2 \to_{k_4} P + E; \\
P &\to_{k_{deg}}; \qquad \to_{k_{prod}} S.
\end{aligned} \qquad (7.2)$$

The corresponding sCCP program is a straightforward extension of the previous one:

$$2\_enz\_reaction(k_1, k_{-1}, k_2, k_3, k_{-3}, k_4, S, E, ES, P) :\!-$$
reaction$(k_1, [S, E], [C_1])$ || reaction$(k_{-1}, [C_1], [E, S])$ || reaction$(k_2, [C_1], [E, P])$ ||
reaction$(k_3, [S, C_1], [C_2])$ || reaction$(k_{-3}, [C_2], [C_1, S])$ || reaction$(k_4, [C_2], [E, P])$

while the rest of the coding is entirely similar to the previous case.

Also in this case a comparison with the reaction obtained with the computed Hill coefficient

$$S \mapsto_{K,V_0,2}^{E} P; \qquad P \to_{k_{deg}}; \qquad \to_{k_{prod}} S,$$

can be easily carried out. This kinetics, however, can be used only under some assumptions that can be found in Section 3.3.2. Notice that the Hill's exponent corresponds exactly to the degree of cooperativity of the enzyme, 2 in this case. Generalization to enzymatic reaction of higher order cooperativity is straightforward. In Figure 7.2, we compare two simulations, one carried out with the full model and one carried out with the reduced model, using the corresponding Hill kinetics. As we can see, the two trajectories are almost coincident.

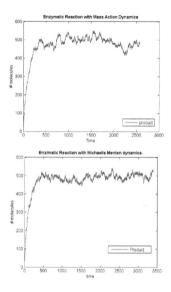

Figure 7.1: (**top**) Mass Action dynamics for an enzymatic reaction. The graph shows the time evolution of the product P. Rates used in the simulation are $k_1 = 0.1$, $k_{-1} = 0.001$, $k_2 = 0.5$, $k_{deg} = 0.01$, $k_{prod} = 5$. Enzyme molecules E are never degraded (though they can be in the complex status), and initial value is set to $E = 10$. Starting value for S is 100, while for P is zero. Notice that the rate of complexation of E and S into ES and the dissociation rate of ES into E and P are much bigger than the dissociation rate of ES into E and S. This implies that almost all the molecules of E will be found in the complexed form. (**bottom**) Michaelis-Menten dynamics for an enzymatic reaction. The graph shows the time evolution of the product P. Rates k_{deg} and k_{prod} are the same as above, whilst $K = 5.1$ and $V_0 = 5$. These last values are derived from mass action rates in the standard way, i.e. $K = \frac{K_2 + k_{-1}}{k_1}$ and $V_0 = k_2 E_0$, where E_0 is the starting quantity of enzyme E, see Section 3.3.1 for a derivation of these expressions. Notice that the time spawn by this second temporal series is longer than the first one, despite the fact that simulations lasted the same number of elementary steps (of the labeled transition system of sCCP). This is because the product formation in the Michaelis-Menten dynamics model is a one step reaction, while in the other system it is a two step reaction (with a possible loop because of the dissociation of ES into E and S).

7.1.2 Example: MAP-Kinase Cascade

A cell is not an isolated system, but it communicates with the external environment using complex mechanisms. In particular, a cell is able to react to external signals, i.e. to signaling proteins (like hormones) present in the proximity of the external membrane. Roughly speaking, this membrane is filled with receptor proteins, that have a part exposed toward the external environment capable of binding with the signaling protein. This binding modifies the structure of the receptor protein, that can now trigger a chain of

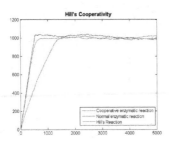

Figure 7.2: In this graph we compare the simulation of the sCCP programs modeling the cooperative enzymatic reaction of equations (7.2) (the two steepest curves). One program uses mass action rates, while the other program uses Hill's kinetic. Basic rates are chosen in order to satisfy the conditions needed to use Hill approximation, see Section 3.3.2, while the parameters for the Hill equation are computed accordingly. As we can see, the two curves are essentially coincident. The third curve represents a normal enzymatic reaction: the cooperativity effect increases the speed of formation of the product.

reactions inside the cell, transmitting the signal straight to the nucleus. In this signaling cascade a predominant part is performed by a family of proteins, called Kinase, that have the capability of phosphorylating other proteins. Phosphorylation is a modification of the protein fold by attaching a phosphorus molecule to a particular amino acid of the protein. One interesting feature of these cascades of reactions is that they are activated only if the external stimulus is strong enough. In addition, the activation of the protein at the end of the chain of reactions (usually an enzyme involved in other regulation activities) is very quick. This behaviour of the final enzyme goes under the name of ultra-sensitivity [133].

$$\text{enz\_reaction}(k_a, k_d, k_r, KKK, E1, KKKE1, KKKS) \parallel$$
$$\text{enz\_reaction}(k_a, k_d, k_r, KKKS, E2, KKKSE2, KKK) \parallel$$
$$\text{enz\_reaction}(k_a, k_d, k_r, KK, KKKS, KKKKKS, KKP) \parallel$$
$$\text{enz\_reaction}(k_a, k_d, k_r, KKP, KKP1, KKPKKP1, KK) \parallel$$
$$\text{enz\_reaction}(k_a, k_d, k_r, KKP, KKKS, KKPKKKS, KKPP) \parallel$$
$$\text{enz\_reaction}(k_a, k_d, k_r, KKPP, KKP1, KKPPKKP1, KKP) \parallel$$
$$\text{enz\_reaction}(k_a, k_d, k_r, K, KKPP, KKKPP, KP) \parallel$$
$$\text{enz\_reaction}(k_a, k_d, k_r, KP, KP1, KPKP1, K) \parallel$$
$$\text{enz\_reaction}(k_a, k_d, k_r, KP, KKPP, KPKKPP, KPP) \parallel$$
$$\text{enz\_reaction}(k_a, k_d, k_r, KPP, KP1, KPPKP1, KP)$$

Table 7.4: sCCP code for the MAP-Kinase signaling cascade. The enz_reaction predicate has been defined in Section 7.1.1. For this example, we set the complexation rates (k_a), the dissociation rates (k_d) and the product formation reaction rates (k_r) equal for all the reactions involved. For the actual values used in the simulation, refer to Figures 7.4 and 7.5.

In Figure 7.3 a particular signaling cascade is shown, involving MAP-Kinase proteins. This cascade has been analyzed using differential equations in [133] and then modeled

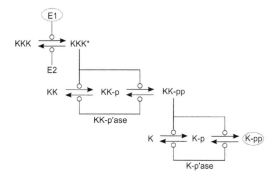

Figure 7.3: Diagram of the MAP-Kinase cascade. The round-headed arrow schematically represents an enzymatic reaction, see Section 7.1.1 for further details.

and simulated in stochastic π-Calculus in [51, 47]. We can see that the external stimulus, here generically represented by the enzyme E_1, triggers a chain of enzymatic reactions. MAPKKK is converted into an active form, called MAPKKK*, that is capable of phosphorylating the protein MAPKK in two different sites. The diphosphorylated version MAPKK-PP of MAPKK is the enzyme stimulating the phosphorylation of another Kinase, i.e. MAPK. Finally, the diphosphorylated version MAPK-PP of MAPK is the output of the cascade.

The sCCP program describing MAP-Kinase cascade is shown in Table 7.4. The program itself is very simple, and it uses the mass action description of an enzymatic reaction (cf. Table 7.2). It basically consists in a list of the reactions involved, put in parallel. The real problem in studying such a system is in the determination of its 30 parameters, corresponding to the basic rates of the reactions involved. In addition, we need to fix a set of initial values for the proteins that respects their usual concentrations in the cell. Following [47], in Figure 7.4 we skip this problem and assign a value of 1.0 to all basic rates, while putting 100 copies of MAPKKK, MAPKK and MAPK, 5 copies of E2, MAPKK-P'ase, and MAPK-P'ase and just 1 copy of the input E1. This simple choice, however, is enough to predict correctly all the expected properties: the MAPK-PP time evolution, in fact, follows a sharp trend, jumping from zero to 100 in a short time. Remarkably, this property is not possessed by MAPKK-PP, the enzyme in the middle of the cascade. Therefore, this switching behaviour exhibited by MAPK-PP is intrinsically connected with the double chain of phosphorylations, and cannot be obtained by a simpler mechanism. Notice that the fact that the network works as expected using an arbitrary set of rates is a good argument in favor of its robustness and resistance to perturbations.

In Figure 7.5, instead, we choose a different set of parameters, as suggested in [133] (cf. the caption of the figure). We also let the input strength vary, in order to see if the activation effect is sensitive to its concentration. As we can see, this is the case: for a low value of the input, no relevant quantity of MAPK-PP is present in the system.

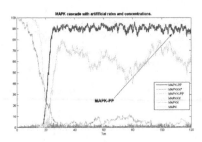

Figure 7.4: Temporal trace for some proteins involved in the MAP-Kinase cascade. Traces were generated simulating the sCCP program of Table 7.4. In this simulation, the rates k_a, k_d, k_r were all set to one. We can notice the sharp increase in the concentration of the output enzyme, MAPK-PP, and its stability in the high expression level. The enzyme MAPKK-PP, the activator of MAPK phosphorylations, instead has a more unstable trend of expression.

7.2   Modeling Gene Regulatory Networks

In a cell, only a subset of genes are expressed at a certain time. Therefore, an important mechanism of the cell is the regulation of gene expression. This is obtained by specific proteins, called *transcription factors*, that bind to the promoter region of genes (the portion of DNA preceding the coding region) in order to enhance or repress their transcription activity. These transcription factors are themselves produced by genes, thus the overall machinery is a networks of genes producing proteins that regulate other genes. The resulting system is highly complex, containing several positive and negative feedback loops, and usually very robust, see Section 3.1.2. Several mathematics techniques have been used to model genetic regulatory networks, see [70] for a survey. However, we focus on a modeling formalism based on logical gates [20], as described in Section 3.1.2.

Specifically, there are three types of gene gates: *nullary gates*, *positive gates* and *negative gates*. Nullary gates represent genes with transcriptional activity, but with no regulation. Positive gates are genes whose transcription rate can be increased by a transcription factor. Finally, negative gates represent genes whose transcription can be inhibited by the binding of a specific protein. At the level of abstraction of [20], the product of a gene gate is not a mRNA molecule, but directly the coded protein. These product proteins are then involved in the regulation activity of the same or of other genes and can also be degraded.

We propose now an encoding of gene gates within sCCP framework, in the spirit of Table 7.1. Proteins are measurable entities, thus they are encoded as stream variables; gene gates, instead, are logical control entities and they are encoded as agents. The degradation of proteins is modeled by the reaction agent of Table 7.2. In Table 7.5 we present the sCCP agents associated to gene gates. A nullary gate simply increases the quantity of the protein it produces at a certain specified rate. Positive gates, instead, can produce their coded protein at the basic rate or they can enter in an enhanced state where production happens at an higher rate. Entrance in this excited state happens at a

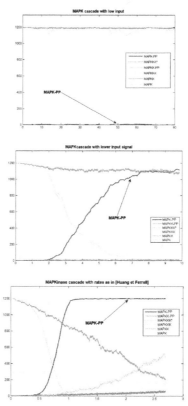

Figure 7.5: Comparison of the temporal evolution of the MAP-Kinase cascade for different concentrations of the enzyme MAPKKK. As argued in [133], this is equivalent to the variation of the input signal E1. Rates are equal for all reactions, and have the following values: $k_a = 1$, $k_d = 150$, $k_r = 150$. This corresponds to a Michaelis-Menten rate of 300 for all the enzymatic reactions. The initial quantity of MAPKK and MAPK is set to 1200, the initial quantity of phosphatase MAPK-P'ase is set to 120, the initial quantity of other phosphatase and the enzyme E2 is set to 5, and the initial quantity of E1 is 1. (**top**) The initial quantity of MAPKKK is 3. We can see that there is no sensible production of MAPK-PP. (**middle**) The initial quantity of MAPKKK is 30. Enzyme MAPK-PP is produced but its trend is not sharp, as expected. (**bottom**) The initial quantity of MAPKKK is 300. The system behaves as expected. We can see that the increase in the concentration of MAPK-PP is very sharp, while MAPKK-PP grows very slowly in comparison.

rate proportional to the quantity of transcription factors present in the system. Negative gates behave similarly to positive ones, with the only difference that they can enter an inhibited state instead of an enhanced one. After some time, the inhibited gate returns to its normal status. A specific gene, generally, can be regulated by more than transcription factor. This can be obtained by composing in parallel the different gene gates.

$$\text{null\_gate}(k_p, X) : -$$
$$\text{tell}_{k_p}(X = X + 1).\text{null\_gate}(k_p, X)$$

$$\text{pos\_gate}(k_p, k_e, k_f, X, Y) : -$$
$$\text{tell}_{k_p}(X = X + 1).\text{pos\_gate}(k_p, k_e, k_f, X, Y)$$
$$+\text{ask}_{r(k_e, Y)}(true).\text{tell}_{k_f}(X = X + 1).\text{pos\_gate}(k_p, k_e, k_f, X, Y)$$

$$\text{neg\_gate}(k_p, k_i, k_d, X, Y) : -$$
$$\text{tell}_{k_p}(X = X + 1).\text{neg\_gate}(k_p, k_i, k_d, X, Y)$$
$$+\text{ask}_{r(k_i, Y)}(true).\text{ask}_{k_d}(true).\text{neg\_gate}(k_p, k_i, k_d, X, Y)$$

$$\text{where } r(k, Y) = k \cdot Y.$$

Table 7.5: Scheme of the translation of gene gates into sCCP programs. The null gate is modeled as a process continuously producing new copies of the associated protein, at a fixed rate k_p. The negative gate is modeled as a process that can either produce a new protein or enter in an repressed state due to the binding of the repressor. This binding can happen at a rate proportional to the concentration of the repressor. After some time, the repressor unbinds and the gate return in the normal state. The enhancing of activators in the pos gate, instead, is modeled here in an "hit and go" fashion. The enhancer can hit the gate and make it produce a protein at an higher rate than usual. The hitting rate is proportional to the number of molecules of the stimulating protein.

7.2.1   Example: Bistable Circuit

The first example, taken from [20], is a gene network composed by two negative gates repressing each other, see Figure 7.6. The sCCP model for this simple network comprehends two negative gates: the first producing protein A and repressed by protein B, the second producing protein B and repressed by protein A. In addition, there are the degradation reactions for proteins A and B. This network is bistable: only one of the two proteins is expressed. If the initial concentrations of A and B are zero, then the stochastic fluctuations happening at the beginning of the simulations decide which of the two fix points will be chosen. In Figure 7.6 we show one possible outcome of the system, starting with zero molecules of A and B. In this case, protein A wins the competition. Notice that the high sensitivity of this system makes it unsuitable for life.

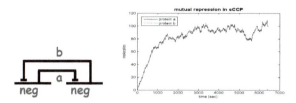

Figure 7.6: Bistable circuit. (**left**) Diagram of gene gates involved. (**right**) Time evolution of the circuit. The negative gates have the same rates, set as follows: basic production rate is 0.1 (k_p in Table 7.5), degradation rate of proteins is 0.0001, inhibition rate (k_i) is 1 and inhibition delay rate (k_d) is 0.0001. Both proteins have an initial value of zero. This graph is one of the two possible outcomes of this bistable network. In the other the roles of the two proteins are inverted.

7.2.2 Example: Repressilator

The repressilator [85] is a synthetic biochemical clock composed of three genes expressing three different proteins, **tetR**, λ**cI**, **LacI**, that have a regulatory function in each other's gene expression. In particular, protein **tetR** inhibits the expression of protein λ**cI**, while protein λ**cI** represses the gene producing protein **LacI** and, finally, protein **LacI** is a repressor for protein **tetR**. The expected behavior is an oscillation of the concentrations of the tree proteins with a constant frequency.

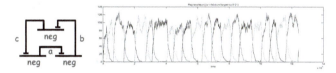

Figure 7.7: Repressilator. (**left**) Diagram of gene gates involved. (**right**) Time evolution of the circuit. The negative gates have the same rates, set as follows: basic production rate is 0.1 (k_p in Table 7.5), degradation rate of proteins is 0.0001, inhibition rate (k_i) is 1 and inhibition delay rate (k_d) is 0.0001. All proteins have an initial value of zero. The time evolution of the repressilator is stable: all simulation traces show this oscillatory behaviour. However, the oscillations among different traces usually are out of phase, and the frequency of the oscillatory pattern varies within the same trace. Remarkably, the average trend of the three proteins shows no oscillation at all, see Chapter 9 for further details.

The model we present here is extracted from [20], and it is constituted by three negative gene gates repressing each other in cycle (see Figure 7.7). The result of a simulation of the sCCP program is shown in Figure 7.7, where the oscillatory behaviour is manifest. In [20] it is shown that the oscillatory behaviour is stable w.r.t. changes in parameters. Interestingly, some models of the repressilator using differential equations do not show this form of stability. More comments on the differences between continuous and discrete models of repressilator can be found in Chapter 9.

7.2.3   Modeling the Circadian Clock

In this section we provide, as a further example, the model of a system containing regulatory mechanism both at the level of genes and at the level of proteins. The system is schematically shown in Figure 7.8. It is a simplified model of the machinery involved in the circadian rhythm of living beings. In fact, this simple network is present in a wide range of species, from bacteria to humans. The circadian rhythm is a typical mechanism responding to environmental stimuli, in this case the periodic change between light and dark during a day. Basically, it is a clock, expressing a protein periodically with a stable period. This periodic behaviour, to be of some use, must be stable and resistant to both external and internal noise. Here with internal noise we refer to the stochastic fluctuations observable in the concentrations of proteins. The model presented here is taken from [234], a paper focused on the study of the resistance to noise of this system. Interestingly, they showed that the stochastic fluctuations make the oscillatory behaviour even more resistant. Our aim, instead, is that of showing how a system like this can be modeled in an extremely compact way, once we have at our disposal the libraries of Sections 7.1 and 7.2.

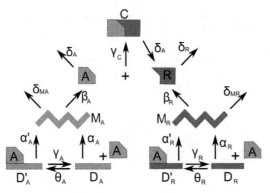

Figure 7.8: Biochemical network for the regulatory system of the circadian rhythm. The figure is taken from [234], like numerical values of rates. Rates are set as follows: $\alpha_A = 50$, $\alpha'_A = 500$, $\alpha_R = 0.01$, $\alpha'_R = 50$, $\beta_A = 50$, $\beta_R = 5$, $\delta_{MA} = 10$, $\delta_{MR} = 0.5$, $\delta_A = 1$, $\delta_R = 0.2$, $\gamma_A = 1$, $\gamma_R = 1$, $\gamma_C = 2$, $\theta_A = 50$, $\theta_R = 100$.

The system is composed by two genes, one expressing an activator protein A, the other producing a repressor protein R. The generation of a protein is depicted here in more detail than in Section 7.2, as the transcription phase of DNA into mRNA and the translation phase of mRNA into the protein are both modeled explicitly. Protein A is an enhancer for both genes, meaning that it regulates positively their expression. Repressor R, instead, can capture protein A, forming the complex AR and making A inactive. Proteins A and R are degraded at a specific rate (see the caption of Figure 7.8 for more details about the numerical values), but R can be degraded only if it is not in the complexed form, while A can be degraded in any form. Notice that the regulation activity of A is modeled by an explicit binding to the gene, which remains stimulated until A unbinds. This mechanism is slightly different from the positive gate described in Section 7.2, but the code can be

$$
\begin{aligned}
&\text{p\_gate}(\alpha_A, \alpha'_A, \gamma_A, \theta_A, M_A, A) \parallel \\
&\text{p\_gate}(\alpha_R, \alpha'_R, \gamma_R, \theta_R, M_R, A) \parallel \\
&\quad \text{reaction}(\beta_A, [M_A], [A]) \parallel \\
&\quad \text{reaction}(\delta_{MA}, [M_A], []) \parallel \\
&\quad \text{reaction}(\beta_R, [M_R], [R]) \parallel \\
&\quad \text{reaction}(\delta_{MR}, [M_R], []) \parallel \\
&\quad \text{reaction}(\gamma_C, [A, R], [AR]) \parallel \\
&\quad \text{reaction}(\delta_A, [AR], [R]) \parallel \\
&\quad \text{reaction}(\delta_A, [A], []) \parallel \\
&\quad \text{reaction}(\delta_R, [R], [])
\end{aligned}
$$

Table 7.6: sCCP program for the circadian rhythm regulation system of Figure 7.8. The agents used have been defined in the previous sections. The first four reaction agents model the translation of mRNA into the coded protein and its degradation. Then we have complex formation, and the degradation of R and A.

adapted in a straightforward manner (we simply need to define two states for the gene: bound and free, cf. below).

The code of the sCCP program modeling the system is shown in Table 7.6; it makes use of the basic agents defined previously, a part from p_gate, which is a redefinition of the positive gate of Table 7.5, taking explicitly into account the binding/unbinding of the enhancer:

$$
\begin{aligned}
&\text{p\_gate}(K_p, K_e, K_b, K_u, P, E) :\text{-} \\
&\quad \text{p\_gate\_off}(K_p, K_e, K_b, K_u, P, E).
\end{aligned}
$$

$$
\begin{aligned}
&\text{p\_gate\_off}(K_p, K_e, K_b, K_u, P, E) :\text{-} \\
&\quad\quad \text{tell}_{K_p}(P = P + 1).\text{p\_gate\_off}(K_p, K_e, K_b, K_u, P, E) \\
&+ \quad \text{ask}_{r_{ma}(K_b, E)}(E > 0).\text{p\_gate\_on}(K_p, K_e, K_b, K_u, P, E)
\end{aligned}
$$

$$
\begin{aligned}
&\text{p\_gate\_on}(K_p, K_e, K_b, K_u, P, E) :\text{-} \\
&\quad\quad \text{tell}_{K_e}(P = P + 1).\text{p\_gate\_on}(K_p, K_e, K_b, K_u, P, E) \\
&+ \quad \text{ask}_{K_u}(\text{true}).\text{p\_gate\_off}(K_p, K_e, K_b, K_u, P, E)
\end{aligned}
$$

In Figure 7.9 (top) we show the evolution of proteins A and R in a numerical simulation performed with the interpreter of the language. As we can see, they oscillate periodically and the length of the period is remarkably stable. Figure 7.9 (bottom), instead, shows what happens if we replace the bind/unbind model of the gene gate with the "hit and go" code of Section 7.2 (where the enhancer do not bind to the gene, but rather puts it into a stimulated state that makes the gene produce *only the next protein* quicker). The result is dramatic, the periodic behaviour is lost and the system behaves in a chaotic way.

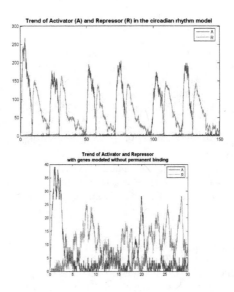

Figure 7.9: Time evolution for circadian rhythm model in sCCP. (top) The figure refers to the system described in Figure 7.8, with parameters described in the caption of the figure. We can see the regularity of the period of the oscillations. (bottom) The graph shows the time evolution for the model where the process governing the gene is the pos_gate described in Table 7.5. The periodic behaviour, with this simple modification, is irremediably lost.

7.3 Encoding core κ-calculus

In Chapter 3 we argued that a protein can be identified by its interface, composed of binding sites and phosphorylation switches. This intuition has been exploited to define κ-calculus [65, 62, 64], having complexation and activation of sites as its primitive operations, see Section 3.2.4. We consider here the core version of the calculus, as presented in [65]. In this calculus, each protein is described by a unique name and by sets of sites, which can be active, inactive or bound. Sites are of two types: binding sites and activation sites; the first ones are involved in complexation activities, while the second ones correspond to the phosphorylation sites. Summarizing, a protein can be identified with $A(\rho, \sigma, \beta)$, where ρ, σ, β are multisets of sites, respectively active, inactive and bound (we can ask that sites of β are coupled with sites they are bound to).

Complexes are composed combining together proteins by the $\odot$ operator, hence a complex is something like $A_1(\rho_1, \sigma_1, \beta_1) \odot \ldots \odot A_n(\rho_n, \sigma_n, \beta_n)$. Rules are of two different kinds, complexation and activation. Complexation rules are of the form

$$A_1(\rho_1 + \tau_1, \sigma_1, \beta_1), \ldots, A_n(\rho_n + \tau_n, \sigma_n, \beta_n) \longrightarrow A_1(\rho_1, \sigma_1, \beta_1 + \tau_1') \odot \ldots \odot A_n(\rho_n, \sigma_n, \beta_n + \tau_n'),$$

binding together proteins at specific sites, while activation rules are

$$A_1(\rho_1, \sigma_1, \beta_1), \ldots, A_n(\rho_n, \sigma_n, \beta_n) \longrightarrow A_1'(\rho_1', \sigma_1', \beta_1), \ldots, A_n'(\rho_n', \sigma_n', \beta_n),$$

switching sites from inactive to active and vice versa. Both kind of rules are local, and can involve proteins contained in bigger complexes. The first rule, in particular, generates a complex combining together all super-complexes.

The calculus presented in [65] is non-deterministic, but it has no stochastic ingredient. A stochastic version seems soon constructed by assigning a basic rate to each rule, and computing its global rule by mass action, i.e. by counting the number of its possible applications. However, determining this number is not an easy task, due to the fact that proteins belong to complexes, fact increasing the combinatorial complexity of the problem. Consider, for instance, the following complexation rule:

$$A(\rho + \{s_1, t_1\}, \sigma, \beta), A(\rho' + \{s_2\}, \sigma', \beta'), A(\rho'' + \{t_2\}, \sigma', \beta') \longrightarrow$$
$$A(\rho, \sigma, \beta + \{s_1|s_2, t_1|t_2\}) \odot A(\rho' + \{s_2\}, \sigma', \beta' + \{s_2|s_1\}) \odot A(\rho'' + \{t_2\}, \sigma', \beta' + \{t_2|t_1\}).$$

To count how many times such a rule can be applied in a specific solution (multiset of proteins and protein complexes), we must find how many A proteins there are with $\{s_1, t_1\}$ among its active sites, how many A proteins there are with $\{s_2\} \subseteq \rho$ and how many A proteins with $\{t_2\} \subseteq \rho$. However, an A protein may well have all the elements $\{s_1, t_1, s_2, t_2\}$ as active sites, though we can use it only once in the application of the rule. It is clear that we must keep track of all possible cases, and the combinatoric of this rule involving only three proteins is already complicated. Of course things go wilder as the number of proteins in the left hand side of a rule grows.

A possible simplification, that enables easier computations of the stochastic rate, is to allow only rules having at most two proteins in the left-hand side. This assumption can also be justified biologically, as all reactions involving more than two molecules are believed to happen in nature as a chain of bimolecular reactions, cf. [99]. For a bimolecular reaction, say a complexation, the most difficult case is when two proteins of the same kind are involved, but using different sites, as in the following rule:

$$A(\rho + \{s_1\}, \sigma, \beta), A(\rho' + \{s_2\}, \sigma', \beta') \longrightarrow$$
$$A(\rho, \sigma, \beta + \{s_1|s_2\}) \odot A(\rho', \sigma', \beta' + \{s_2|s_1\}). \tag{7.3}$$

In this case, if we denote with $n_1 = N(A, \{s_1\})$, $n_2 = N(A, \{s_2\})$, $n_{12} = N(A, \{s_1, s_2\})$ the number of proteins with s_1, s_2 and s_1, s_2 among their active site respectively, the number of possible applications of the rule is

$$(n_1 - n_{12})(n_2 - n_{12}) + (n_1 - n_{12})n_{12} + (n_2 - n_{12})n_{12} + n_{12}(n_{12} - 1) = n_1 n_2 - n_{12}, \quad (7.4)$$

where we count separately the interactions between proteins having just one or both the requested sites.

With this simplification in mind, we can start the presentation of an encoding of κ-calculus in sCCP. This can be seen as a first step of a direct encoding of Kohn maps, see [145]. First of all, to encode proteins in sCCP, we represent them by a name (identified by an unique constant) and three lists, one for active sites, one for hidden sites and one for bindings. A natural representation of a complex is as a list of proteins; therefore we use the predicate

$$\texttt{complex}(Names, ActiveSitesList, HiddenSitesList, BindingsList, X),$$

where $Names$ is a list of the names of the proteins forming the complex, $ActiveSitesList$ is a list of lists of active sites, one for each protein in $Names$, and $HiddenSitesList$ and $BindingsList$ are defined similarly. Finally, X is the number of complexes of this form present on the solution. We also require that there is an ordering among names of proteins, and an ordering among names of sites, in such a way that each complex has an unique representation as a predicate[12]. In order to compute correctly the rates of reactions, we need to store the number of proteins whose list of sites contains specific singletons and specific pairs of sites. Sites are of two kinds: binding sites and phosphorylation switches. We need to count only active binding sites, while we need to consider both active and inactive phosphorylation switches (marking them with an 1 or a 0 to distinguish between the two forms). This information can be stored in predicates like $\texttt{number}(P, SiteList, X)$, where P is the name of the protein, $SiteList$ contains one or two sites of interest and X is the number of proteins with these characteristics. It's important to remark that these proteins can be free in the solution, or part of bigger complexes.

The encoding of κ-calculus will be reaction-centric, associating a $\texttt{complexation}$ or an $\texttt{activation}$ agent to each reaction of the system, as one would expect from the usual practice with sCCP. As a matter of fact, this is in line also with the rule-centric approach of κ-calculus. However, in order to manage correctly complexes, we find convenient to introduce an agent for each complex type. Agent of this kind are all instantiations of the same procedure, $\texttt{complex\_manager}$. A definition of such agents can be found in Tables 7.7, 7.8 and 7.9; part of the details of the operations involved is hidden in the constraints used.

---

[1]If a complex contains two or more proteins with the same name, we can extend the ordering to multisets of sites, so as to remove ambiguity in the ordering.

[2]This condition allow to use in a simple way the unification of Prolog to check if a particular complex is present or not.

As a first simplification, the variables A_i, A_i', and C in the code represent proteins and complexes, hiding the internal complexity of these objects (they are represented by 4 lists).

$\text{complexation}(A_1, A_2, A_1', A_2', ID, R)$:-
$\qquad\text{ask}_{\lambda_{reaction}(R, A_1, A_2)}(\texttt{is\_present}(A_1, A_2))$.
$\qquad\text{tell}_\infty(\texttt{signal\_reaction}(ID))$.
$\qquad\text{ask}_F(\texttt{reply}(ID))$.
$\qquad\exists_C\ (\text{tell}_\infty(\texttt{build\_final\_complex}(ID, A_1, A_2, A_1', A_2', C))$.
$\qquad\qquad\text{tell}_\infty(\texttt{adjust\_counters}(A_1, A_2, A_1', A_2', C))$.
$\qquad\qquad\text{tell}_\infty(\texttt{add\_unit}(C))$.
$\qquad\qquad\quad\text{ask}_F(\texttt{is\_present}(C))$.
$\qquad\qquad\qquad\text{complexation}(A_1, A_2, A_1', A_2', ID, R)$.
$\qquad\qquad+\ \text{ask}_F(\texttt{is\_not\_present}(C))$.
$\qquad\qquad\qquad(\ \text{complexation}(A_1, A_2, A_1', A_2', ID, R)$
$\qquad\qquad\qquad\|\ \text{complex\_manager}(C)\ )\ )$

Table 7.7: Definition of complexation agent of sCCP encoding of κ-calculus.

$\text{activation}(A_1, A_2, A_1', A_2', ID, R)$:-
$\qquad\text{ask}_{\lambda_{reaction}(R, A_1, A_2)}(\texttt{is\_present}(A_1, A_2))$.
$\qquad\text{tell}_\infty(\texttt{signal\_reaction}(ID))$.
$\qquad\text{ask}_F(\texttt{reply}(ID))$.
$\qquad\exists_{C_1, C_2}\ (\text{tell}_\infty(\texttt{modify\_sites}(ID, A_1, A_2, A_1', A_2', C_1, C_2))$.
$\qquad\qquad\text{tell}_\infty(\texttt{adjust\_counters}(A_1, A_2, A_1', A_2', C_1, C_2))$.
$\qquad\qquad\text{tell}_\infty(\texttt{add\_unit}(C_1, C_2))$.
$\qquad\qquad\quad\text{ask}_F(\texttt{is\_present}(C_1) \wedge \texttt{is\_present}(C_2))$.
$\qquad\qquad\qquad\text{activation}(A_1, A_2, A_1', A_2', ID)$.
$\qquad\qquad+\ \text{ask}_F(\texttt{is\_not\_present}(C_1) \wedge \texttt{is\_present}(C_2))$.
$\qquad\qquad\qquad(\ \text{activation}(A_1, A_2, A_1', A_2', ID, R)$
$\qquad\qquad\qquad\|\ \text{complex\_manager}(C_1)\ )$
$\qquad\qquad+\ \text{ask}_F(\texttt{is\_present}(C_1) \wedge \texttt{is\_not\_present}(C_2))$.
$\qquad\qquad\qquad(\ \text{activation}(A_1, A_2, A_1', A_2', ID, R)$
$\qquad\qquad\qquad\|\ \text{complex\_manager}(C_2)\ )$
$\qquad\qquad+\ \text{ask}_F(\texttt{is\_not\_present}(C_1) \wedge \texttt{is\_not\_present}(C_2))$.
$\qquad\qquad\qquad(\ \text{activation}(A_1, A_2, A_1', A_2', ID, R)$
$\qquad\qquad\qquad\|\ \text{complex\_manager}(C_1)$
$\qquad\qquad\qquad\|\ \text{complex\_manager}(C_2)\ )\ )$

Table 7.8: Definition of activation agent of sCCP encoding of κ-calculus.

The encoding we propose is similar in spirit to the first one presented in [65], mapping κ-calculus into π-calculus. In fact, we take a reaction-centric view, associating an agent to each reaction of the system, be it complexation or an activation. In addition, we need an agent to manage each complex type, together with a mechanism to create such an agent when a new complex type has been formed. We need agents managing complexes because complex_manager(C) :-

$\sum_{ID \in Reactions}$ ($\quad$ ask$_{\lambda_c(A_1,C,R_2)}$(has_happened$(ID) \wedge$
$$is\_in\_complex(A_1, C, R_2)).$$
$$tell_\infty(react(R_1, ID)).$$
$$complex\_manager(C)$$
$+ \quad$ ask$_{\lambda_c(A_2,C,R_1)}$(has_happened$(ID) \wedge$
$$is\_in\_complex(A_2, C, R_1)).$$
$$tell_\infty(react(R_2, ID)).$$
$$complex\_manager(C))$$

$+ \quad$ ask$_F$(has_participated_in_last_reaction$(R_1, R_2)).$
$$tell_\infty(decrease\_element(C, R_1, R_2).$$
$$(ask_F(greater\_than\_0(C)).complex\_manager(C)$$
$$+ \ ask_F(equal\_to\_0(C)).\mathbf{0})$$

$\lambda_c(A, C, R) = F \cdot number\ of\ occurences\ of\ A\ in\ copies\ of\ C,\ given\ R.$

Table 7.9: Definition of complex manager agent of sCCP encoding of κ-calculus.

rules of κ-calculus apply locally to proteins part of bigger complexes. Hence, in addition of choosing, with the correct rate, one rule to apply, we need to choose the specific protein involved in the reaction, identifying also the complex it belongs to. This is fundamental, as these local rules modify globally the complex containing the reagents. This is the factor of complexity of the model, forcing to introduce a protocol of synchronization between reaction agents and managers of complexes.

The complexation agent, presented in Table 7.7, takes as input the preconditions and post-conditions of its rule, i.e. proteins A_1, A_2, A_1', A_2', an id of the rule, needed to synchronize with other processes, and the basic rate R of the reaction. With its first instruction, it checks if there are the proteins A_1 and A_2 in the solution (with requested active sites), competing in a stochastic race. The rate of this ask instruction counts the number of possible applications of the rule, according to the formula (7.4), and then multiplies it by the basic rate R, passed as parameter. Note that we are using a mass action kinetics.

If the agent wins the race condition, it signals to other processes that the reaction indeed happened. This is obtained by telling constraint signal_reaction(ID), setting to 1 the two stream variables X, previously 0, stored in predicates reaction(ID, i, X, Y, Z), with $i = 1, 2$. It also sets to 0 variables Y, that is used to exchange information about complexes to merge. In addition, it also sets to ID the stream variable X stored in the predicate last_reaction(X). Then, the process waits that the managers of complexes compete for deciding the actual complexes involved. Once this has been done, the stream variables X in reaction(ID, i, X, Y, Z) will be equal to two, and (at least one of) the variables Y will be instantiated with the complexes involved in the reaction. Finally, the variables Z will store the position of the molecule A_i of the complex indicated by Y taking part in the reaction. When managers have finished to update these variables, the instruction asking reply(ID) is active, and the complexation agent proceeds by building the complex created by the reaction (build_final_complex). This is done by substituting A_i' to A_i in the complexes stored in variables Y of reaction(ID, i, X, Y, Z). The variable

Z stores the correct position of the A_i to replace. If the reaction involves two proteins belonging to the same complex, just one of the Y variables will be different from 0 (but both Z will indicate the correct position)[3]. The final complex constructed is stored in the local variable C. The process then increases the counter of the number of molecules of complex C (add_unit(C)), stored in the predicate complex(C, X), and adjusts the value of the variables counting the number of active sites, stored in number($P, SiteList, X$) and used to compute correctly stochastic rates. Finally, the agent controls if the new complex build was already present in the solution or not. If not, it spawns a new complex_manager agent for it, and then calls itself recursively. Note that all ask instructions except the first one have rate F. This rate is a very big number, several order of magnitudes bigger than rates of reactions. This guarantees that, with very high probability, this instruction will be executed before all other instructions with much smaller rates. Intuitively, one would like to have ask guards with infinite rate, but this is not permitted by our language, in order to guarantee the confluence property for the instantaneous transition, see Chapter 6 for further details.

The functioning of the agent activation, see Table 7.8, managing activation reactions, is similar to the agent complexation, with the only difference that it produces two final complexes, instead of one. This forces to introduce straightforward modifications in order to manage correctly the spawning of complex_manager agents.

The last class of agents needed are those managing the different complex types, i.e. complex_manager agents. These agents are defined as a big summation over all possible reactions; for each reaction, there are two branches in the summation, one dealing with the possibility that the complex managed contains the first element of the reaction, the other with the possibility of containing the second element. This summation has also a final branch, restoring the initial conditions of the agent, whenever it was involved in a reaction. Each branch managing a reaction has an ask guard, checking if the reaction ID has just happened (with constraint has_happened(ID), looking if stream variables X of predicates reaction(ID, i, X, Y, Z) are equal to one) and controlling if the protein A_i involved in the reaction ID is a member of the complex (with the correct configuration of active sites). This is realized by the constraint is_in_complex(A_i, C, R_j). R_j, $j = 3 - i$, is a local variable needed to tackle the case in which another element of the complex has been chosen as the protein A_j. If this is the case, R_j will be equal to the position in C of the chosen R_j,[4] otherwise it will be equal to zero.[5] The function giving the rate of this ask instruction is defined at the bottom of Table 7.9. It counts the number of possible agents that can contribute to reaction ID, by inspecting the complex C and counting how many occurrences of A there are and multiplying this number by the number of copies of C in the solution (stored in the stream variable X of complex(C)); if one A has already been chosen as the other protein involved in the reaction (this is specified by R, if it points to an agent of type A), the number is modified accordingly, by subtracting one. The resulting integer is multiplied by the fast rate F, in order to execute this instruction immediately. Note that, after a reaction ID has been executed, we have a "fast" race condition among all those complex_manager agents that can participate to ID, each counted with the

---

[3]Lists defining the complex are always reordered after each operation modifying them.

[4]We also need to keep track of which copy of the complex C in the solution has been chosen. This can be obtained by substituting variables R_i with pairs of variables, or by a simple encoding of such pair as an integer. We omit this detail hereafter.

[5]We skip over the correct initialization of variables R_i.

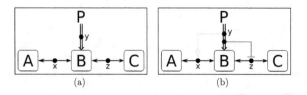

Figure 7.10: Two very simple MIMs

correct multiplicity. Once a guard has been overcome, the agent tells which is the protein chosen by the constraint $\text{react}(R_i, ID)$. This agent changes also the value of variables in $\text{reaction}(ID, i, X, Y, Z)$, by setting $X = 2$ (element chosen), Y to C (only if R_j is equal to zero - meaning that the complex is not used twice), and Z to the position of A_i in C. Then, the agent calls itself recursively again. The last branch of the big external summation is active whenever the constraint $\text{has\_participated\_in\_last\_reaction}$ is enabled by the store. This happens if both proteins involved in the last reaction have been chosen,[6] and if at least one of R_i is different from zero. When this branch is active, the R_i are set back to a value of zero, and the number of units of the complex C is decreases by one or two, depending on R_i. The last test checks if all the copies of C have been consumed or not. If that happens, then the agent complex_manager dies, otherwise it calls itself again.

7.4 Encoding Molecular Interaction Maps

In this section we define an encoding in sCCP of Molecular Interaction Maps (MIM), a graphical notation to describe biochemical networks defined by Kohn [147]. MIMs have been roughly introduced in Section 3.2.4, and they will be presented in more detail hereafter. The crucial ingredient of the encoding will be the *implicit* representation of complexes and reactions. Essentially, molecular complexes will be represented by graphs, modified locally by reactions. All this information will be stored in the constraint store in suitable predicates. This section owes a lot to the previous one: it can be seen as a generalization of the encoding of κ-calculus capable of dealing with more general situations. Part of the following material appeared in [31].

7.4.1 Molecular Interaction Maps

Molecular Interaction Maps, or Kohn Maps, have been already introduced in an informal way in Section 3.2.4. We provide now a more detailed, yet incomplete, presentation. The interested reader is referred to [147] for more details. A MIM is essentially a graph, where nodes correspond to different (basic) molecular species, linked by different kinds of connecting lines, see Figure 7.10 for an example. The notation follows few general principles: to keep the diagram compact an *elementary molecular species*, like A, B or C in Figure 7.10(a), generally occur in *only one place* on a map. Different interactions between molecular species, instead, are distinguished by different lines and arrowheads; for

---

[6]The id of last reaction happened can be extracted from the predicate last_reaction (ID).

instance, the double line in Figure 7.10(a) represents a covalent modification of B (in this case a *phosphorylation*, i.e. the attachment of a phosphate group at a specific place in the protein), while the single double-barbed arrows denote complexation operations. *Complex molecular species*, or simply complexes, are created as a consequence of interactions and are indicated by small circles on the corresponding interaction line; e.g. in Figure 7.10(a) x represents the complex $A : B$, the result of a complexation between A and B, while y represents the phosphorylated B molecule (denoted hereafter with pB).

In general, there are two types of interaction lines: *reactions* and *contingencies*. The former operate on molecular species, the latter on reactions or other contingencies. The red line with a T-shaped end of Figure 7.10(b) is an *inhibition* line; it states that phosphorylated B (indicated by y) cannot bind to C, (in fact the line terminates on the barbed arrow connecting B and C). Another contingency symbol of Figure 7.10(b) is the arrow with a bar preceding its empty arrowhead, terminating in x. This line represents a *requirement*: B must be phosphorylated in order to bind to A. Note that multiple nodes on an interaction line represent exactly the same molecular species. In Figure 7.11 we list a subset of arrows, taken from [147].

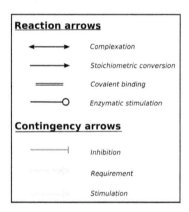

Figure 7.11: A restricted list of reaction and contingency arrows of MIMs.

In order to fully explain the differences between Figure 7.10(a) and Figure 7.10(b), we need to introduce the *interpretations* of MIMs. The MIM notation, in fact, can be equipped with two different interpretations: *explicit* and *combinatorial*.

In the explicit interpretation, an interaction line applies only to the molecular species directly connected to it. Looking again at Figure 7.10(a), we can say that B can bind to A (forming the complex $A : B$) or to C (forming the complex $B : C$) but there is no way the complex $A : B : C$ will be formed. Moreover, if B is phosphorylated it cannot bind neither to A nor to C.

In the combinatorial interpretation, an interaction line represents a functional connection between domains or sites that (unless otherwise indicated) is independent of the modifications or bindings of the directly interacting species. In this way the map of Figure 7.10(a) states that A can bind to B, independently of the state of B. For instance, B could be phosphorylated and thus forming the $A : pB$ complex, or it could be bound to C, resulting in the $A : B : C$ complex.

The first interpretation needs every interaction to be explicitly defined, in a "what is depicted is what happens" fashion, and this requires the use of many reaction lines. On the other hand, the combinatorial interpretation takes the opposite point of view, i.e. "what is not explicitly forbidden does happen". In this case the behavior can be specialized by means of contingency lines.[7]

Complexes	Figure 7.10(a)		Figure 7.10(b)	
	Expl.	Comb.	Expl.	Comb.
A:B	✓	✓		
B:C	✓	✓	✓	✓
A:pB		✓	✓	✓
pB:C		✓		
A:B:C		✓		
A:pB:C		✓		

Table 7.10: Different interpretations of MIMs of Figure 7.10

In Table 7.10 we summarize the differences between the two interpretations. While the two interpretations of the map of Figure 7.10(a) are different, the map of Figure 7.10(b) has exactly the same behavior under both interpretation, thanks to the use of inhibition and requirement symbols. Moreover, the combinatorial interpretation represents implicitly a large number of molecules: a single complexation arrow usually handles, on both sides, a big set of reactants. Consider, for example, the arrow connecting A to B in Figure 7.10(a); in the combinatorial interpretation it represents four reactions, namely the complexation of A with B, pB, $B : C$ and $pB : C$.

In principle, it is always possible to create an explicit MIM with the same behaviors of a combinatorial MIM, introducing more reaction arrows and contingencies. Explicit MIMs can be easily translated into a set of ODEs or into an explicit stochastic model for computer simulation, see [144]. Unfortunately, expliciting a combinatorial MIM is a non-trivial job, due to the combinatorial explosion of reaction arrows and contingencies needed in the map. In addition, an explicit ODE-based (or stochastic) simulation, like the one described in [144], requires to consider all possible molecular complexes that can be generated in the system as reactants or products of some reaction. However, at a given time, usually only a small subset of these complexes is present, hence the effort needed to generate explicitly this large number of complexes is largely unmotivated.

Our goal is precisely to define a simulation operating directly at the level of the combinatorial interpretation of MIMs, hence this is the interpretation we will consider in the following. The advantages of this choice are clear: during each stage of a simulation, we need to represent only the complexes present in the system at that time. Moreover, models can be defined using the compact notation of combinatorial MIMs, hence the resulting map is usually smaller, thereby easier to build and understand.

---

[7]There is also a third layer of interpretation, called *heuristic* in [146], used mainly to structure and organize available knowledge of biological systems. In this interpretation, all interactions possible in the combinatorial case and forbidden in the explicit one are left unspecified, thus representing incomplete knowledge.

Removing ambiguity from MIMs

The biggest obstacle towards the definition of an implicit encoding of MIMs is, unfortunately, in the MIM notation itself. In fact, *the combinatorial interpretation is far from being a rigorous semantic*: combinatorial MIMs are intrinsically ambiguous, as there are many cases in which the exact behavior of the map is not defined. Moreover, some arrows in the MIM notation are just "syntactic sugar", as their behavior can be defined in terms of other arrows. To tackle these problems, we have defined a set of *graph rewriting rules*, disambiguating some cases in what we deem a biologically plausible way and removing redundant arrows. This can be seen as the first step towards the definition of a formal semantic for MIMs. We discuss some examples in the following.

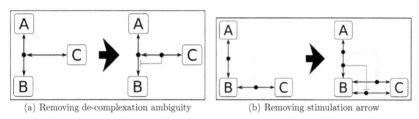

(a) Removing de-complexation ambiguity   (b) Removing stimulation arrow

Figure 7.12: An example of graph rewriting rules for MIMS

The MIM of Figure 7.12(a) represents a situation in which A can bind to B, forming a complex $(A : B)$ that can bind to C. Writing this in the standard, self-explanatory, chemical notation, we obtain $A + B \to A : B$ and $(A : B) + C \to A : B : C$. Biologically, every complex formed this way can, in principle, be de-complexed, meaning that the map of Figure 7.12(a) describes also the inverse reactions $A : B : C \to (A : B) + C$ and $A : B \to A + B$. It is not clear, however, if the link between A and B in $A : B : C$ could be broken or not. We assume that this is not possible, because the map states that C can bind only to $A : B$ and not to A nor B alone, thus it plausibly describes a biological mechanism in which C binds to a modified structure containing A and B, a structure that cannot disappear while bound to C.[8]

In order to make this choice explicit, we add an inhibition symbol from the node representing the $A : B : C$ complex to the $A : B$ complexation arrow, thus preventing de-complexation of $A : B$ when C is bound. This is an example of a very simple graph rewriting rule in order to eliminate ambiguity.

Figure 7.12(b) left, shows a MIM where the $A : B$ complex *stimulates* the $B : C$ bond (stimulation is represented with an arrow with empty arrowhead). The intended meaning of stimulation is that B is more likely to bind to C when it is complexed with A. We can see this mechanism in an alternative way, using two complexation arrows between B and C, one representing the slower reaction and the other one representing the faster reaction. These arrows need to be mutually exclusive, see Figure 7.12(b) right. This second rewriting rule reduces the number of symbols of maps, simplifying the encoding in sCCP. Therefore, the two examples given are representative of two classes of rewriting rules: the first tackling ambiguity, the second simplifying the notation.

---

[8]Note that different behaviors can be obtained changing the topology of the map. For instance, if C binds only to A instead of binding to $A : B$, the complex $A : B : C$ can be broken also in $(A : C) + B$.

7.4.2   Encoding MIMs in sCCP

In this section we present the basic ideas for the direct simulation of MIM in sCCP. The starting point is the definition of a suitable representation of molecules and complexes. Actually, with respect to other process algebras like π-calculus [194], sCCP offers a crucial ingredient in this direction, namely the presence of the constraint store. In fact, the store is customizable and every kind of information can be represented by the use of suitable constraints, i.e. logical predicates. Manipulating and reasoning on such information can be performed in a logic programming style [214]. These ingredients make the constraint store an extremely flexible tool that can be naturally used to represent the data structures needed to operate on MIMs.

The idea to simulate MIMs is simple: we operate directly on the graphical representation. Of course, a MIM contains all possible interactions of the system, hence the graphical representation used in the simulation must be specialized to single molecules and complexes. Complexes, in particular, can be seen as *graphs*, with *nodes representing the basic molecules* (i.e. the proteins) of the complex, and with *edges representing the chemical bonds* tying them together. Such graphs will be referred in the following as *complex-graphs*.

In the discussion on biochemical reactions of Section 3.1.1, we discussed how molecular species can be defined by the collection of their interaction sites. In our encoding, these sites are represented by *ports*, having the following properties:

- each port (or better, *port-type*) is the terminating point of one single arrow in the MIM[9];

- each port-type is characterized by an unique identifier, called *port_type_id*, and by a boolean variable, INH_p, storing the state of the port. In fact, each port can be active ($INH_p = false$), meaning that it can take part to the corresponding reaction, or inhibited ($INH_p = true$) by biological mechanisms specified in the map;

Molecular-types correspond to nodes of the MIM or to points in the middle of an arrow (i.e., terminating points of reaction arrows). They consist of an unique identifier, *molecular_type_id*, of a list of port-types, implicitly determining all the possible reactions the molecule can be involved into, and of a list of contingencies starting from it (we present the treatment of contingencies at the end of the section). For instance, in Figure 7.10(a), A and x nodes define two distinct molecular-types.

Each graph of a complex can contain, in principle, several instances of the same molecular-type, just think of the case in which two copies of the same molecule are bound together (the so-called homodimers). In order to distinguish among these different copies, we introduce an unambiguous naming system inside each complex, enumerating each node of a complex-graph with an integer, local to that complex, called *mol_id*. Moreover, each different complex graph that can be constructed according to the prescription of the MIM, identifies a *complex-type*; complex-types are also given an unique id, *complex_type_id*, assigned at run-time whenever a new complex-type is created.

Complex-graphs and molecular-types can be easily represented in sCCP. In fact, we just need to store all the characterizing information in suitable logical predicates, listed at

---

[9]This condition can always be made true by the application of suitable rewriting rules to the map.

> molecular_type($molecular\_type\_id$,
> $port\_list, contingency\_list$)
> node($molecular\_type\_id$, mol_id)
> edge([mol_id_1, $port\_type\_id_1$],
> [mol_id_2, $port\_type\_id_2$])
> complex_type($complex\_type\_id$, $node\_list$,
> $edge\_list$, $contingency\_list$)
>
> complex_number($complex\_type\_id$, X)
> port_number($port\_type\_id$, X)

Table 7.11: Predicates describing MIM structures and counting their occurrences.

the beginning of Table 7.11. Note that each molecule in a complex-type is unambiguously identified by the pair ($complex\_type\_id, mol\_id$).[10] Such pairs are called *coordinates* of the molecule.

Another important class of predicates, crucial for the run-time engine, are those counting the number of objects of a certain type. Specifically, at run-time we need to count how many complexes we have for each different complex-type (predicate complex_number), and how many active ports we have, for any port-type (predicate port_number). The variable X used in these predicates is a stream variable.

The reason for updating the number of ports or complexes at run-time, lies in the definition of the stochastic model for the simulation of MIMs. We adopt a classical approach, defining the speed of a reaction according to the *principle of mass action* [102]: the speed of each reaction is proportional to the quantity of each reactant. In our encoding, each reaction involves one or two port-types, hence its speed will be proportional to the number of active instances of such port-types.

The simulation algorithm of such stochastic model is based on the celebrated Gillespie algorithm [101, 102], extended in order to manage all the additional information of graphs and complexes. Essentially, we can see it as a loop composed of 4 basic steps:

1. choose the next reaction to execute;

2. choose the reactants;

3. create the products;

4. apply contingency rules to products.

We give now some details of its sCCP implementation. The choice of the next reaction can be seen as a stochastic race among all the enabled reactions. In sCCP, this effect is obtained associating an agent to each reaction arrow of the molecular interaction map, called *reaction agent*. For instance, in Table 7.12 we show the agent dealing with complexation. This agent tries to execute at a rate defined according to the principle of

---

[10] $molecular\_type\_id$ can be recovered from mol_id and the predicate node($molecular\_type\_id$, mol_id).

complexation($port\_id_1$,$port\_id_2$, $rate$) :-
    $\exists$ $complex\_id_1$,$complex\_id_2$,mol_id_1,mol_id_2,$nodes$,
      $edges$,$contingencies$,$new\_complex\_id$.
    ask$_{[rate \cdot \#port\_id_1 \cdot \#port\_id_2]}$(
      are_greater_than_zero($port\_id_1$,$port\_id_2$)
      $\wedge$ **are_reactions_unlocked**).
    tell$_\infty$(**lock_reactions**).
    tell$_\infty$(**activate_port_manager**($port\_id_1$)).
    ask$_F$(**is_port_manager_done**($port\_id_1$)).
    tell$_\infty$(**activate_port_manager**($port\_id_2$)).
    ask$_F$(**is_port_manager_done**($port\_id_2$)).
    tell$_\infty$(**get_chosen_complex**($port\_id_1$,$complex\_id_1$,mol_id_1)).
    tell$_\infty$(**get_chosen_complex**($port\_id_2$,$complex\_id_2$,mol_id_2)).
    tell$_\infty$(**build_complex**($complex\_id_1$,mol_id_1,$complex\_id_2$,
      mol_id_2,$nodes$,$edges$,$contingencies$)).
    tell$_\infty$(**add_complex**($nodes$,$edges$,$contingencies$,
      $new\_complex\_id$)).
    tell$_\infty$(**update_numbers**($complex\_id_1$,$complex\_id_2$,
      $new\_complex\_id$,$port\_id_1$,$port\_id_2$)).
    tell$_\infty$(**apply_contingencies**($new\_complex\_id$)).
    tell$_\infty$(**unlock_reactions**).
    complexation($port\_id_1$,$port\_id_2$, $rate$)

        port_manager($port\_id$) :-
          $\exists$ $complex\_id$,mol_id.
          ask$_F$(**is_port_manager_active**($port\_id$)).
          tell$_\infty$(**choose_complex**($port\_id$,$complex\_id$,mol_id)).
          tell$_\infty$(**complex_chosen**($port\_id$,$complex\_id$,mol_id)).
          tell$_\infty$(**port_manager_done**($port\_id$)).
          port_manager($port\_id$)

Table 7.12: Two sCCP agents used in the simulation of MIMs.

mass action (ask$_{[rate \cdot \#port\_id_1 \cdot \#port\_id_2]}$). If the agent wins the competition, it gets control of the whole system (**lock_reactions**) and then it activates two auxiliary agents in order to identify the reactants actually involved (**activate_port_manager**). In fact, a reaction involves two complexes with available ports of the required port-type. However, as different complex-types can have such ports available, we must identify the ones really involved in the reaction. Essentially, we need to pick, with uniform probability, one complex among all those having an active port of the required type. This operation is performed by *port managers* (see again Table 7.12): there is one agent for each port-type, keeping an updated list of all complex types containing active ports of its type and choosing one of them upon request from a reaction agent (**choose_complex**). Port agents keep the information about active ports in a list stored in a dedicated predicate, whose elements are pairs ($complex\_type\_id$, mol_id) identifying the molecule in a complex-type containing the port (we omitted these details from the code of Table 7.12).

When reactants have been chosen by port agents, the reaction agent generates the products of the reaction (`build_complex`). For instance, in case of a complexation reaction, the agent has to merge two complexes into one, adding nodes and edges to the complex description and marking as bound the ports involved in the reaction (i.e. removing them from lists of the corresponding port agents).

If, after these operations, a new (i.e. not present in the system) complex-type is obtained, then all the necessary predicates are added to the constraint store. Otherwise, the value's variable counting the number of complexes of the generated type is increased, while those of the two reactants are decreased (`add_complex`,`update_numbers`). Checking if a complex-type is already present is, in fact, a problem of graph isomorphism. For the subclass of complex-graphs, this problem is quadratic in their description, thanks to the fact that ports are unique for each edge, see [207] for further details.

The last point of the simulation algorithm consists in the application of contingencies (`apply_contingencies`). These are the inhibition and requirement arrows, briefly introduced in Section 7.4.1. To grasp the rationale behind their implementation, consider again Figure 7.10(b). The inhibition arrow from y (the head of the contingency rule) to A-B complexation line (the tail) was used to forbid complexation between phosphorylated B and A. This essentially means that, if B is phosphorylated (i.e., the corresponding edge e is in the complex description), then B cannot bind to A, and so the port p_{B-A} connecting B to A, must become inactive.

This example suggests that contingencies are nothing but logical implications of the form:

$$\text{IF (a set of edges } E \text{ is in the complex)}$$
$$\text{THEN (some ports must become active/inactive)}$$

Rules of this kind are stored in each molecular-type descriptor, so that each complex-type has associated a set of contingency rules potentially applicable.

When a new complex type is created in a reaction, then the reaction agent checks what rules among those listed in the complex can be applied, and it modifies accordingly the value of INH_p of ports and their associated global counters, like the predicate `port_number` and the lists used by port agents.

7.4.3 Experimental Results

In this section we present some preliminary tests of the framework just presented. To test the encoding of MIMs in sCCP, we simply defined the templates of all agents and predicates needed; in this way, a MIM can be described simply by the collection of agents associated to reactions and ports and by the description of molecular types. Then, a specific program is fed to the Prolog interpreter of sCCP (cf. Sections 6.5), and simulations are performed. Unfortunately, the interpreter based on Prolog lacks computational efficiency. In the future, we plan to write a more efficient implementation of the whole language.

The example we consider is a very simple MIM taken from [144], where Kohn first introduced the MIM notation in order to study the mammalian G1/S cell-cycle phase transition. In that paper, Kohn studied this system using an evolutionary perspective: he started from a small subnetwork and extended it to include further details at each step. Here we consider the simplest system, composed by three proteins. One is the

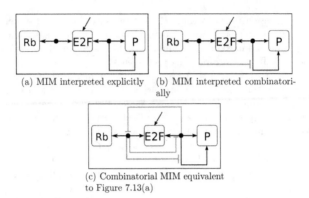

(a) MIM interpreted explicitly   (b) MIM interpreted combinatori-
                                       ally

(c) Combinatorial MIM equivalent
to Figure 7.13(a)

Figure 7.13: Molecular Interaction Maps for a simple subnetwork of the G1/S phase
transition network.

transcription factor $E2F$, involved in the regulation of several genes active in the replica-
tion of DNA (the S phase of the cell cycle). The second is a protein Rb, belonging to the
Retinoblastoma family, inhibiting $E2F$ activity when complexed to it. The third is a Protease P involved in the degradation of $E2F$. The corresponding map, as taken from [144],
is shown in Figure 7.13(a). As we can see, there is also an external feeding of $E2K$. The
problem with this map is that in [144] Kohn interpreted it explicitly. This means that
the bindings of $E2F$ are exclusive. In the combinatorial interpretation, instead, these two
bindings can coexist in the same complex. However, in both maps the degradation of
$E2F$ can happen when $E2F$ is bound only to protease P. This is a choice that we make
explicit introducing a suitable inhibition arrow, see Figure 7.13(b). If we want to obtain
the same behavior of the explicit interpretation, we need to make the bindings exclusive:
this can be obtained simply adding two inhibition symbols in the map, ending up with
Figure 7.13(c).

The differences between the two interpretations are not confined to the topology of the
maps, but they obviously propagate to the dynamical behavior. In Figure 7.14, we show
the simulation of the map of Figure 7.13(b) (Figure 7.14(a)) and of that of Figure 7.13(c)
(Figure 7.14(b)). As we can see, the dynamics and the stable values of the complex
Rb:E2F are different. In fact, in the less restrictive system of Figure 7.13(b), part of
the $Rb : E2F$ complex is also bound to protease P, hence the stationary value of pure
$Rb : E2F$ is lower.

This example, despite its simplicity, shows two things: the feasibility of our encoding,
which is able to simulate the maps implicitly, and the fact that defining possible interaction
in combinatorial MIMs is a matter of forbidding unwanted or unrealistic behaviors.

7.4.4   Final Discussion

Molecular Interaction maps are an implicit notation: each edge in such a diagram represents a set of reactions potentially very large. Our sCCP-simulation, instead of generating the full list of reactions, is able to simulate the map implicitly, generating only

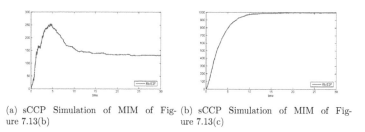

(a) sCCP Simulation of MIM of Figure 7.13(b)

(b) sCCP Simulation of MIM of Figure 7.13(c)

Figure 7.14: sCCP simulation of MIMs of Figures 7.13(b) and 7.13(c). Parameters are as in [144].

those complexes that are actually present in the system at run-time. This is achieved using a graph-based representation of complexes, so that new complexes are dynamically constructed merging and splitting other complex-graphs.

The choice of sCCP as a language to describe (implicitly) such maps has several advantages. First of all, the power of constraints allows to represent and reason directly on the graph-based representation of complexes, separating this description from the definition of agents performing the simulation. Another important motivation is that the model built in such way is compositional w.r.t. the addition of new edges (and nodes) in a MIM. In fact, agents are associated to each reaction arrow of the map (plus an agent to each port type). Hence, adding arrows requires just to put in parallel an agent of the corresponding type (plus the port managers for the new ports). The addition of new basic molecular species, instead, requires the addition in the store of the predicates describing them. Finally, the sCCP description of a MIM is proportional in size to the number of symbols needed to write the map itself, thus taming the combinatorial explosion of possible reactions.

We turn now to compare our encoding of MIMs with other related works aiming at modeling and simulating bio-regulatory networks.

Kohn himself, in [144], uses the explicit version of MIMs to generate the associated mass action ODEs. Of course, this requires all possible reagents to be defined, giving rise to the combinatorial explosion discussed in Section 7.4.1.

Another approaches, closer to ours, are β-binders [54]. β-binders [196] are a modification of stochastic π-calculus [194] in which π-processes are contained into boxes, having interaction capabilities exposed in their surface (the so called binders). Processes in different boxes can communicate through typed channels having an high affinity provided by an external function. Boxes can also be split and merged at run-time. MIMs can be described as β-binders [54] using the same basic assumptions we made in Section 7.4.2: molecules are represented as boxes and characterized by the collection of their interaction sites (i.e. of their binders). However, this encoding differs from ours as it follows the principles put forward in [203]: each basic molecule present in the system will be described by a dedicated box. The description of complexes, instead, is stored in the external environment. Clearly, different basic molecules have different descriptions in terms of boxes (they differ in the inner π-processes and in their binders), hence the addition of new interaction capabilities requires the modification of all boxes involved. Our encoding, instead, obtains a kind of compositionality on the edges of the map. Another substantial difference with β-binders is that we are not developing a new language, but we are simply programming an existing one, without even exploiting all its features. For instance, the functional rates can be used, as in [36, 26], to encode chemical kinetics different from mass action, like Michaelis-Menten one, resulting in a simplification of models (we need one reaction instead of three).

An even more direct approach to model MIM is given by κ-calculus, see Section 3.2.4, which is very similar to the spirit of our encoding. However, the advantage of using sCCP w.r.t κ-calculus, to our advice, resides in its programmability. In fact, the encoding of MIMs has been realized simply by the definition of a suitable library of constraint predicates and agents. Nothing prevents to integrate it with other functionalities, for instance the management of dynamic compartments (a feature missing also in MIMs).

Finally, we need to compare this encoding to that of the previous section. As a matter of fact, they are quite similar, the main differences being:

- `complex_manager` agents are replaced by `port_manager` agents. The advantage is that port managers need not to be spawned at run-time, hence there is an a-priori control of the size of the program that needs to be executed.

- The mechanism to check isomorphism of two complexes is different: in Section 7.3 we rely on an encoding of complexes as strings, while in this section we use directly the graph representation.

We can also observe how both encodings make use of the fast rate mechanism to simulate guarded instantaneous actions. This is an approximation pointing to a limit of the current version of sCCP: we should introduce more complex instantaneous actions, providing them with a priority mechanism inducing discrete probability distributions on instantaneous traces that can be integrated smoothly in the CTMC-based semantics of sCCP.

7.5   Model Checking sCCP using PRISM

In this section we focus on the problem of verifying whether models written in sCCP satisfy certain specifications, formalized using stochastic continuous logic (CSL, see Section 2.5.3 of Chapter 2). Instead of defining and implementing a model checker tailored on sCCP, we decided to use the well developed PRISM [155], an open source software freely available on the net [1]. PRISM specifies stochastic models, among which CTMC, using a proprietary language, see Section 2.5.5 for a quick introduction. Therefore, we need to define an encoding of sCCP into PRISM. PRISM language, in order to guarantee the finiteness of the CTMC, has a series of restriction that must be dealt with. Basically, it allows only a finite and fixed number of processes in parallel, hence no new process can be forked at run-time. In addition, these processes, called *modules*, can modify the value of a set of variables, either local or global. The instructions are executed asynchronously and they are guarded by a conjunction of conditions on the variables. All the instructions whose guards are active compete in a race condition; they can also contain inner non-determinism in form of a summation, see again Section 2.5.5. Finally, variables in PRISM take value on integers, and their domain is always bounded, hence finite.

sCCP is more expressive than PRISM, as processes can be forked and killed at run-time and the structure of the space of variables enjoys the richness of a constraint store.

$$Program = D.N$$

$$D = \varepsilon \mid D.D \mid p(\mathbf{x}) : -A$$

$$\pi = \text{tell}_\lambda(c) \mid \text{ask}_\lambda(c)$$
$$M = \pi.G \mid M + M$$
$$G = \mathbf{0} \mid \text{tell}_\infty(c) \mid p(\mathbf{y}) \mid M \mid \exists_x G \mid G.G$$
$$A = \mathbf{0} \mid \text{tell}_\infty(c) \mid M \mid \exists_x A \mid A.A$$
$$N = A \parallel A \mid N$$

Table 7.13: Syntax of the sequential version of sCCP; the restrictions involve the use of the parallel operator, that now can be used to combine agents only at the network level.

Therefore, in order to define a reasonable mapping, we need to restrict it to a subset that is compatible with the assumptions of the model checker.

The first set of restrictions must be made at the level of the constraint store. First of all, we allow only the use of stream variables, posting on them just assignments in the form of equality constraints. The constraints of guards, in addition, are restricted to equality and inequality constraints on arithmetic expressions combining variables. In the following, we denote with $\mathcal{U}$ the set of assignments that can be told and with $\mathcal{G}$ the set of conditions that can be asked. Moreover, the domains of our variables must be bounded; the programmer has the task to control the boundary conditions with *ad-hoc* instructions in order to avoid overflow and underflow errors.

Restrictions must also involve the syntax of the language in order to have a fixed number of *sequential agents* in parallel, where a sequential agent is any agent not containing parallel operators. The resulting sub-language can be syntactically defined as in Table 7.13.

A final restriction regards the names of variables used in the definition of procedures and the names of local variables. Specifically, we require these names to be all distinct; moreover, a procedure definition can be used only within one component of the network: if one procedure is used in more than one component, we need to define several syntactic copies of it. A straightforward consequence of this restriction is that global variables can be shared only among agents constituting the initial configuration of the system (i.e. those forming the initial network N, see Table 7.13).

7.5.1 Translating sCCP into PRISM

We are now ready to define a translation procedure that maps each program of (restricted) sCCP into a corresponding program of PRISM. We explain this method via an example; a formal treatment is demanded to Chapter 9.

Consider the following sCCP program:

```
RW(X) :-
        ask₁(X > 0).tell∞(X = X − 1).RW(X)
```

$+$ $\text{ask}_1(X < M).\text{tell}_\infty(X = X + 1).\text{RW(X)}$
$+$ $\text{ask}_{f(X)}(true).(\quad \text{ask}_1(X > 1).\text{tell}_\infty(X = X - 2).\text{RW(X)}$
$\qquad\qquad + \text{ask}_1(X < M - 1).\text{tell}_\infty(X = X + 2).\text{RW(X)} \;)$

$$f(X) = \frac{K}{\left(\frac{M}{2} - X\right)^2 + 1}$$

This agent performs a variant of random walk on the single integer variable X. With rate 1, it increases or decreases the value of the variable by one unit. Notice that the ask guards guarantee that the value of X is always contained in $[0, M]$, hence X is bounded as requested. The final branch of the sum makes the agent enter in an "excited state", where X can be increased or decreased by 2 units, always with rate 1. The entrance in this state happens at a rate $f(X)$, inversely proportional to the distance of X from the midpoint of its definition interval, i.e $\frac{M}{2}$.

The translation procedure is better explained and visualized using syntactic trees, where each node either contains an instruction or is a summation node. No parallel composition nodes exist, as we are dealing with sequential agents. We also drop the existential operator defining local variables, to recover this information in a following phase. The basic idea of the translation is to collapse all transitions with infinite rate following a stochastic-timed one; to do this, we first move the information about guards, updates and rates of transitions from nodes to edges, and then collapse the edges. We deal with recursive calls introducing cycles in the syntactic tree, effectively constructing a compact form of the labeled transition system of the agent. The syntactic tree for the agent of our example is the following:

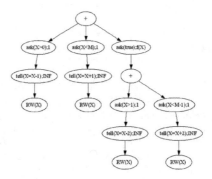

The first step of the procedure works by moving some information from nodes to edges. In particular, we label the incoming edge of a node with a triple where the first element is a finite set of guards of $\mathcal{G}$, the second element is a set of assignments taken from $\mathcal{U}$, and the third element is the rate function of that transition. Nodes containing ask instructions contribute just with guards and rates, while tell agents contribute with updates and rates (consistency check for assignments is trivial). We also remove labels from nodes corresponding to ask or tell. Summation nodes and procedure calls are left untouched, though incoming arcs of procedure calls are labeled by $(\emptyset, \emptyset, \infty)$. In the following diagram, an empty label is denoted by "*".

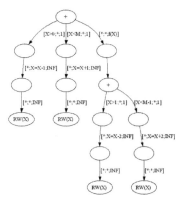

The following step consists in removing the nodes corresponding to recursive calls. This can be achieved in two ways, depending on the presence or absence of a copy of the syntactic tree of the called procedure in the graph under manipulation. If no copies of the tree of the called procedure are present in the graph, we replace the calling node with one copy of such tree (with the transformations of first step already performed). Moreover, we add to the label of the incoming edge of the procedure call node the assignments performed by the mechanism linking variables to formal parameters of the called procedure. If, instead, there already exists a copy of the tree of called procedure in the graph we are managing, we simply redirect the incoming edge of the node calling the procedure to the root of the copy of such tree, adding to the edge the corresponding linking assignments. In this way, we can introduce cycles in the graph, like in our example.

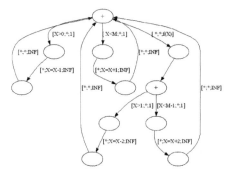

The next sequence of steps removes all nodes having a single outgoing edge with infinite rate and a single incoming edge. If our initial agent is in stochastic normal form (see Chapter 6), we can be guaranteed that all nodes with more than one incoming edge have just stochastic edges (edges whose rate labeling them is not ∞) exiting from them. Therefore, we are removing all nodes not corresponding to stochastic summations, i.e those nodes defining a sequence of instantaneous and deterministic steps without non-determinism and stochastic duration. When these nodes are removed, their entering and exiting edge are merged, and the corresponding labels are joined together: if (g_1, u_1, r_1)

and (g_2, u_2, r_2) are such labels, the resulting one will be $(g_1 \wedge g_2, u_1 \wedge u_2, \min\{r_1, r_2\})$. Note that at least one between r_1 and r_2 will be equal to ∞, hence we are not mixing together rates of non-instantaneous transitions. At the end of this procedure, all edges will have a stochastic rate less than infinity. At some point during the execution of this step, the diagram of our example looks like this:

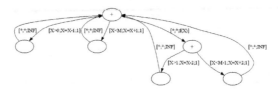

After merging all possible edges, we clean all labels of remaining nodes, assigning an integer value to each of them, identifying the state of the system. The resulting graph for the agent of the example is the following:

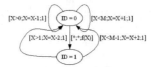

In general, the resulting graphs have nodes labeled by integers and edges labeled by guards that must be enabled to perform the transition, by the updates on the variables of the system induced by the transition, and by the rate of the transition. We call this kind of graphs *reduced transition systems* (RTS). In Chapter 9, we give a formal definition and derivation of RTS for a slightly more restricted language, essentially forbidding local variables. An extension to this case can be easily given. In a RTS we have all the information needed to write a PRISM program. The restrictions imposed to the constraint store and to admissible constraints in ask and tell guarantee that their form is compatible with PRISM (they are assignments of the form $X = f(X, Y, Z, \ldots)$ and comparisons like $Y \Diamond g(X, Y, Z, \ldots)$, with f, g arithmetic functions and $\Diamond \in \{<, \leq, =, \geq, >\}$).

The PRISM program is written in the following way: for each agent constituting the initial network of the system, we define a PRISM module. Global variables shared by agents are defined as global variables of PRISM, while local variables declared with $\exists$ are defined as local variables in each module. We suppose to know the bounds on the domain of each variable. A dedicated local variable named ID is defined for each module. We also need to fix the initial value of each variable. A module has as many instructions as are the edges of the RTS of the agent; if an edge connects node j to node k and has label $(g_1 \wedge \ldots \wedge g_n, u_1 \wedge \ldots \wedge u_m, r)$, the corresponding PRISM instruction is

```
[] (ID=j)&(g1)&...&(gn) -> r:(u1)&...&(un)&(ID' = k);
```

We also merge instructions having the same guards (hence exiting from the same node) using the non-deterministic choice of PRISM, see Section 2.5.5. The resulting PRISM module for the sCCP agent of the example is

```
module RW
    [] (ID = 0) & (X > 0)       ->     1 : (X' = X-1);
    [] (ID = 0) & (X < M)       ->     1 : (X' = X+1);
    [] (ID = 0)                 -> f(X) : (ID' = 1) ;
    [] (ID = 1) & (X > 1)       ->     1 : (X' = X-2) & (ID' = 0);
    [] (ID = 1) & (X < M-1)     ->     1 : (X' = X+2) & (ID' = 0);
endmodule
```

The complete PRISM program, obtained converting the sCCP program containing just the RW(X) agent, simply needs to declare the variable X as global, with domain $[0, M]$ and initial value $\frac{M}{2}$, using the following instruction:

```
global X : [0..M] init M/2;
```

7.5.2 Examples of Model Checking Queries in PRISM

In this section we show how to use PRISM to extract information about the dynamic behaviour of sCCP programs. The first step of this analysis is obviously to convert them into an equivalent PRISM program, that need to be loaded into the PRISM model checker. Once this is done, we can start checking some queries written in Continuous Stochastic Logic, see Section 2.5.3. Actually, PRISM supports a richer class of queries, like computation of transient and steady-state probabilities, and computation of average quantities and expected time to reach specific sets of states.

In the following, we show some of the possible analysis via examples, the first one dealing with the agent RW(X) of the previous section. This system performs a variant of a random walk in one dimension, where the average step is bigger near the middle value of the domain, i.e. $\frac{M}{2}$, say $M = 200$. Moreover, the domain of X is bounded, and when the system reaches the boundary, it is "bumped away" from it. From basic theory on stochastic processes (see, for instance, [181]), we know that the variance of a random walk is unbounded, growing as the square root of the number of steps performed. Hence, we expect that the system reaches the boundary with probability one, starting from $\frac{M}{2}$. One way to formalize this property in CSL is to say that the probability of eventually reaching the boundary is one:

$$\mathcal{P}_{\geq 1}(\Diamond(X = 0 \vee X = M)).$$

In the specification language supported by PRISM, this property is specified using the characterization of $\Diamond$ via the until operator:

```
P>=1 [true U X=0 | X=M]
```

If we check this property against the PRISM model of the agent RW(X), we obtain a positive answer, as expected.

Stochastic model checking is based on algorithms that compute explicitly the value of the probability of a path formula, and then compare this value with the bound in the $\mathcal{P}$ operator. PRISM can be instructed to exhibit directly the computed probability, using the operator P=?. For instance, we may wonder what is probability of reaching the bound $X = M$ and from it, with probability 1, the bound $X = 0$. We can express this query as

```
P=? [ true U x = M & P>=1 [true U x=0]  ],
```

whose probability, computed by the model checker, is 1.

PRISM is also able to compute steady-state probabilities, and thus answer queries regarding steady-state probability of state formulae. For instance, we may ask what is the long run probability of X being equal to $\frac{M}{2}$:

$$S=? \ [X = MAX/2].$$

The reply to such query is 0.0015. So, despite the fact that $\frac{M}{2}$ is the average value of the system, we do not expect it to be equal to that value very often.

Finally, PRISM is also capable of computing average quantities. In order to do this, we must specify suitable cost functions. For instance, we may set instantaneous costs of transition to zero, and cumulative costs in states to one, so that the cost of a trace coincides with its temporal duration. In addition, PRISM can compute the average value of the costs associated to states at a given time t. In this way, we can compute the average quantity at a certain time of a function of our state variables, like the average quantity of X at time t. This is achieved with the query

$$R=? \ [I=t].$$

PRISM allows the user to perform a number of these queries, varying the parameter t; therefore, we can plot a graph with the trend of the average value of X, see Figure 7.15.

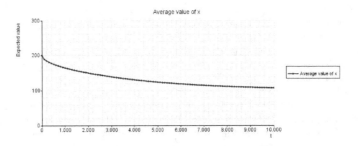

Figure 7.15: Average value of X for agent, RW(X), starting from initial value of M. This graph has been generated by PRISM.

The same machinery can be used to plot graph of values dependent of parameters. For instance, we may ask what is the probability of X dropping below 10, starting from M, within t units of time, and then plot this probability as a function of t. The corresponding query uses the time-bounded until operator:

$$P=? \ [\ true \ U<=t \ X<10 \ ];$$

the graph of the probability is shown in Figure 7.16.

As a second example, we apply model checking to analyze some properties of a simple biological system, i.e. an enzymatic reaction, see Section 7.1.1 for the sCCP model. Essentially, we apply the translation into PRISM both for the mass-action model and the corresponding Michaelis-Menten model. To apply model checking, we must restrict the domains of the variables, so that they become bounded. This is not the case for any biochemical reaction, though we can always choose a very big integer as bound, reachable

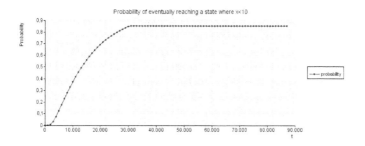

Figure 7.16: Probability of X dropping below 10 within t units of time, for agent RW(X), starting from initial value of M. This graph has been generated by PRISM.

with low probability, so that the behaviour of the model is not altered sensibly. However, performances of the model checker depend on the size of the underlying CTMC, hence variables should have the smallest domain as possible. For this simple reaction, we choose as upper bound of the domain the value of 50, which was never reached by the product P in few runs of the simulation.

The first question we ask to the model checker is with which probability such bound will eventually be reached:

$$P=? \; [\; true \; U \; P=MAX \;].$$

The answer to such query is 1, meaning that if we let the system run for a long time, we expect to see for sure P reaching the upper bound we fixed on its domain. But how much time we have to wait? We tried to compute the average time required to reach a state satisfying $P = MAX$, by setting the suitable cost function. Unfortunately, ever for this simple model (with bounds on domains we have set, the CTMC has less than 30000 states), the model checker failed to answer within one hour. Therefore, we turned to look at the probability of eventually reaching $P = MAX$ within 5000 time units, querying PRISM with

$$P=? \; [\; true \; U<=5000 \; P=MAX \;].$$

This probability is around 0.0225, suggesting that the time to see such an event occur is very big, compared to the time of stabilization of the Markov process. In fact, as we will see in a while, P stabilizes around its average value within 50 units of time.

A second set of questions we feeded into the model checker regarded the steady state behavior. First of all, we asked what was the average value of P in the long run, with

$$R=? \; [\; S \;],$$

obtaining a value of 25. Then we asked what was the steady-state probability of P being equal to 25 (query S=? [P=25],answer 0.08) or contained in the interval [15, 35] (query S=? [P<=35 & P>=15], answer 0.97). Finally, we plotted the time-evolution of the average value of P, with the same method used for the agent RW(X) (i.e, asking R=?[I=t]), obtaining the graph in Figure 7.17.

To see if the Michaelis-Menten and the Mass Action stochastic models for this enzymatic reaction were equivalent, we performed the same experiments about steady-state probabilities and average value of P also for the Michaelis-Menten sCCP agent. PRISM stated that the long run average of P is 25, that the steady-state probability of P being equal to 25 is 0.08, and that the steady-state probability of P belonging to $[15, 35]$ is 0.97. All these values are very close to the corresponding ones for mass action (equal when rounded off). In addition, we also plotted the time-evolution of the average value of P, obtaining the graph of Figure 7.18. Hence, model checking experiments confirm the essential equivalence of these two formulations of an enzymatic reaction.

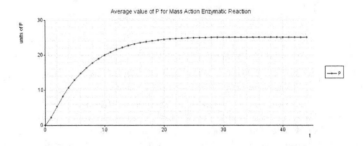

Figure 7.17: Time evolution of the average value of P for the Mass Action sCCP program describing an enzymatic reaction with feeding of S and degradation of P, see Section 7.1.1. Parameters used here are $k_1 = 0.1$, $k_{-1} = 0.01$, $k_2 = 0.5$, $k_s = 2.5$ and $k_d = 0.1$, while initial concentration of the enzyme is $E_0 = 10$.

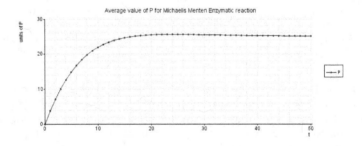

Figure 7.18: Time evolution of the average value of P for the Michaelis-Menten sCCP program describing an enzymatic reaction with feeding of S and degradation of P. Parameters were calculated from those in the caption of Figure 7.17: $K = 5.1$ and $V_{max} = 5$.

7.5.3   Final Comments

Model checking is just one among the analysis techniques applied to stochastic process algebras. For instance, one may study behavioral equivalences like Markovian bisimulation [128], or compute the steady state of a CTMC [128]. An interesting question is whether these methods will lead to any insight while applied to models of biological systems. Model checking, on the other hand, looks more promising, as temporal logic (in its stochastic variant) can express dynamical properties of systems. However, the real issue with model checking is computational, as biological systems have a huge state space that is not tractable with current methods. A possibility to tackle this problem is to define a stochastic form of abstract interpretation, and use it to reduce the dimension of the state space, essentially restricting the set of admissible values of the molecules to a small number of concentration levels, though large enough to guarantee preservation of qualitative behavior. The first step in this direction is to define a theory of stochastic abstract interpretation, possibly capable of measuring a-priori the approximation error. A related work is the probabilistic abstract interpretation [75, 76], dealing with probabilistic programs evolving in discrete time.

Chapter 8

Multi-Agent Protein Structure Prediction in sCCP

The Protein Structure Prediction Problem (PSP), described widely in Chapter 4, is the problem of predicting the 3D *native* conformation of a protein, when the sequence made of 20 kinds of amino acids (or *residues*) is known.

In this Chapter we present an high-level framework for ab-initio prediction of the protein structures using agent-based technologies. In this presentation, we follow the lines of [30, 38], extending the versions of [28, 29]. The predictor is developed by following the architecture for agent-based optimization systems presented by Milano and Roli in [169]. This framework stratifies the agents in different levels, according to their knowledge and their power. Here we have three layers: one containing agents designed to explore the state space, one dealing with agents implementing global strategies and the last one containing cooperation agents.

Each amino acid in the protein is modeled as an independent agent, which has the task of exploring the configuration space. This is accomplished mainly by letting these agents interact and exchange information. These processes operate within a simulated annealing scheme (see Chapter 5), and their moves are guided by the knowledge of the position of surrounding objects. The communication network changes dynamically during the simulation, as agents interact more often with their spatial neighbors. The strategic agents govern the environmental properties and they also coordinate the basic agents activity in order to obtain a more effective exploration strategy of the state space. The cooperative agent, instead, exploits some external knowledge, related to local configurations attainable by a protein, to improve the folding process.

The complexity of this framework seems quite distant from the kind of systems we have modeled in sCCP so far. However, the flexibility of constraints and the concurrency of the language make sCCP adapt to describe also this application. In fact, sCCP can model systems described at different degrees of granularity, incorporating in an easy way all the information that is needed. In this case, we are basically encoding a simulation of the protein folding, based on a Monte Carlo stochastic dynamics regulated by other agents of the model. Moreover, the level of detail required in the specification of this system necessitates of mechanisms for storing and managing spatial coordinates, for computing complex functions and for making stochastic choices depending on these functions. sCCP can effectively deal with all these things, thanks to the generality of the logical language behind the store, the use of the inference machinery of Prolog and the non-constant rates (see Chapter 6).

In this chapter, we focus our attention mainly on the prediction framework, leaving out as many details of the sCCP implementation as possible. Part of these details, in particular the implementation of the Monte Carlo simulation scheme, are demanded to Appendix A.

Due to the high requirements in terms of computational power, we have designed a dedicated multi-thread implementation in C, which is equivalent to the sCCP description, though much faster (see Section 8.2.4).

In the following, we use two different models of proteins. Both represent amino acids as a single center of interaction, coincident with C_α atom, so they differ in the energy function used. The first one has been developed by Micheletti in [71], while the second has been presented in [28]. Both potentials have a low resolution, hence the models predicted are not expected to be very accurate. However, we are mainly concerned with the reliability of the minimization scheme and its stability of finding good minima of the energy, a fact confirmed by experimental results. In fact, more refined potentials can be used, as the framework is modular and independent from the energy function, like the one recently proposed in [91].

The chapter is organized as follows. In Section 8.1 we describe the two energy model employed, while in Section 8.2 we present the Agent-based framework. In Section 8.2.4 we provide some details about the implementation and, finally, in Section 8.3 we discuss the results of some experiments.

8.1   Energy function

The problem of identifying an accurate energy for a simplified representation of the aminoacids is considered very difficult, and there is no general accordance relatively to which one reflects better the physical reality. Consequently, in literature there is a plenty of different energy functions one can choose from. All these models share a common feature: the more accurate are the results, the more complex are the calculations involved.

We present in detail the two energy functions that we used to perform our tests. Comparison of their performances can be found in Section 8.3.

8.1.1   First Energy Model

The first energy model is the one developed by Micheletti in [71]. Each aminoacid is represented by a single center of interaction, identified with the C_α carbon atom. The energy comprises three terms, devoted to interaction, cooperation and chirality. In addition, it contains some hard constraints that forbid non physical configurations. However, our simulations run in a concurrent framework, and each agent (aminoacid) does not generally have at his disposal the most recent information about the position of other aminoacids (cf. Section 8.2.1 for further comments). Hence, using hard constraints can result in a too strict and rigid policy for the movement, making some configurations unreachable and violating the balance equation for the Monte Carlo method used in the exploration of the search space. A possible way out to this problem is to convert the hard constraints into soft ones, i.e. into smooth energy barriers that (heavily) penalize the non-physical configurations.

In the following we describe the energy terms involved in a concise way; further details can be found in [71]. If we indicate with $\mathbf{x}$ the spatial disposition of the aminoacid's chain and with $\mathbf{t}$ their type, the energy can be expressed as

$$E(\mathbf{x}, \mathbf{t}) = \begin{aligned} & E_{coop}(\mathbf{x}, \mathbf{t}) + E_{pairwise}(\mathbf{x}, \mathbf{t}) + \\ & E_{chiral}(\mathbf{x}, \mathbf{t}) + E_{constr}(\mathbf{x}) \end{aligned} \tag{8.1}$$

The *pairwise term* captures the interactions that occur between two aminoacids that are close enough, and its formal expression is

$$E_{pairwise}(\mathbf{x}, \mathbf{t}) = \sum_{i=1}^{n-2} \sum_{j=i+2}^{n} \varepsilon_{i,j} f(r_{i,j}), \tag{8.2}$$

where n is the number of aminoacids in the chain, $r_{i,j}$ is the distance between aminoacid i and j, f is a sigmoidal function and $\varepsilon_{i,j}$ represent the strength of the interaction between the two residues i and j. In particular, the function f models the dependance of the intensity from the distance, and is defined as

$$f(r) = \frac{1}{2} - \frac{1}{2} \tanh(6.5 - r), \tag{8.3}$$

so that it goes rapidly to 0 if $r > 6.5$ and to 1 if $r < 6.5$. The matrix of the interaction weights $(\varepsilon_{i,j})$ is the one developed by Kolinsky in [148].

The *cooperative term* involves four different aminoacids, and it tries to improve the packing of secondary motifs. It has the following expression

$$E_{coop}(\mathbf{x}, \mathbf{t}) = \frac{1}{20} \sum_{i=1}^{n-1} \sum_{j=i+1}^{n} \sum_{a,b} f(r_{i,j}) f(r_{i+a,j+b}) [\varepsilon_{i,j} + \varepsilon_{i+a,j+b}], \tag{8.4}$$

where f is defined by equation (8.3), $a, b \in \{\pm 3, \pm 4\}$, and ε is the same interaction matrix of equation (8.2). This term advantages the situations where the aminoacid i is bonded j, and one aminoacid close to i is bonded to one close to j.

The *chiral term*, instead, is used to favor the formation of helices for some putative segments, identified using knowledge extracted from PDB database. Further details can be found in [71]. Its form is

$$E_{chiral}(\mathbf{x}, \mathbf{t}) = V_0 \sum_{i=1}^{n-3} p_i \left[\frac{1}{2} + \frac{1}{2} \tanh \left(\left(\frac{\chi - 0.7}{\sigma_\chi} \right) \left(\frac{6.0 - r}{\sigma_r} \right) \right) \right], \tag{8.5}$$

where χ is the torsional angle of 4 consecutive aminoacids, r is the distance between the first and the last aminoacids in the segment, V_0 equals 3.0 and p_i is 1 if the segment starting with aminoacid i is a putative helical fragment and 0 otherwise.

In addition to these three energy terms, we have several constraints implemented via energy barriers, which are collected together in the term E_{constr}:

$$E_{constr}(\mathbf{x}) = E_{steric}(\mathbf{x}) + E_{chircst}(\mathbf{x}) + E_{dist}(\mathbf{x}).$$

E_{steric} imposes three steric constraints to the position of C_α and C_β atoms. The position of C_β is calculated from the chain of alpha carbon atoms using the Park and Levitt rule (cf. [189]). If x_i and x_j are the positions of two carbon atoms (either α or β)

for aminoacids i and j, then it must hold that $r = \text{dist}(x_i, x_j) > 3$. This is achieved by means of a Lennard-Jones potential barrier of the form $\left(\frac{3}{r}\right)^6$ if $r < 3$ and 0 otherwise.

The $E_{chircst}$ is similar in spirit to the last term, a part from forbidding regions of the plane (χ, r), where χ and r are defined as in equation (8.5). In particular, the following three regions are penalized: $\chi < 0$ and $r < 7.5$ Å, $r < 4$ Å and $r > 11$ Å.

Finally, the E_{dist} term tries to keep fixed the distance between two consecutive C_α atoms, around the value of 3.8 Å. This is achieved through a parabolic potential of the form $(r - 3.8)^2$.

Penalty terms are weighted experimentally in order to balance their effects with the original energy terms. Actual numerical values can be found in http://www.dimi.uniud.it/dovier/PF/.

8.1.2   Second Energy Model

The second energy model has been developed in [28] to perform the tests on the first version of the simulation engine. However, that version presented several limitations in the way the search space was explored, and this is why we decided to test it again with this more powerful architecture.

In the following we describe this energy function with a certain detail. It is composed by four separate terms, related to *bond distance* (E_b), *bend angle* (E_a), *torsion angle* (E_t), and *contact interaction* (E_c). Like in previous section, we indicate with **x** the vector denoting the position of aminoacids, and with **t** the vector of their types.

Bond distance energy.   For each pair of *consecutive* aminoacids x_i, x_{i+1}, we have a quadratic term

$$E_b(\mathbf{x}, \mathbf{t}) = \sum_{1 \leq i \leq n-1} \left(r(x_i, x_{i+1}) - r_0 \right)^2 \tag{8.6}$$

where $r(x_i, x_{i+1})$ is the distance between the $C\alpha$ of the two aminoacids and r_0 is the typical distance of 3.8 Å between the two $C\alpha$. The fact that the distance is close to 3.8 Å is typically a hard constraint. Here this fact is expressed by the quadratic increase of the energy w.r.t. the distance from equilibrium.

Bend angle energy.   Another term composing the energy function is the bend energy associated to a triplet of consecutive aminoacids. The bend angle β of $C\alpha_i, C\alpha_{i+1}$, and $C\alpha_{i+2}$ is simply the angle formed by the two bonds linking the three carbons (cf. Fig 8.1).

This value is almost constant in every protein and independent from the types of aminoacids involved, but we preferred to run a cycle of profiling and approximation over the PDB using a single class of 20 residues. The resulting profile matches the one described in [90] and presents almost a Gaussian distribution around 120 degrees (precisely, $\beta_2 = 113.7$), plus a small sharp peak around 90 degrees (precisely, $\beta_1 = 88.6$), due to helices. The energy is obtained by applying the opposite logarithm to the distribution function. The formula associated is of the form

$$E_a(\mathbf{x}, \mathbf{t}) = \sum_{i=1}^{n-2} -\log\left(a_1\, e^{-\left(\frac{\beta_i - \beta_1}{\sigma_1}\right)^2} + a_2\, e^{-\left(\frac{\beta_i - \beta_2}{\sigma_2}\right)^2} \right) \tag{8.7}$$

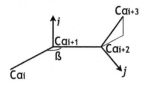

Figure 8.1: Bend angle β

where β_i is the bend angle depending on (x_i, x_{i+1}, x_{i+2}). The angles β_1, β_2 are expressed in radians, a_i are the coefficients, and σ_i are the variances; their values are shown in Figure 8.2. In the same figure the energy contribution using the parameters computed with our statistics is plotted.

i	1	2
a_i	0.995325	0.305612
β_i	1.546907	1.984492
σ_i	0.089444	0.415203

Figure 8.2: Values for bend parameters and Approximated distribution of bend angles

Torsion angle energy. We introduce a heuristic function that we developed by gathering torsional information from the Protein Data Bank [16]. In more detailed simulations (cf. [179]), part of the energy function is related to torsion angles about each bond. Thus, we include in our model some statistics of torsional information to guide and improve the folding.

In detail, four $C\alpha$ atoms ($C\alpha_i, C\alpha_{i+1}, C\alpha_{i+2}$, and $C\alpha_{i+3}$) form a specific torsion angle. This angle is influenced by the type of the aminoacids involved and by their position in the protein. Let $\vec{r}_{12} = C\alpha_{i+1} - C\alpha_i$, $\vec{r}_{23} = C\alpha_{i+2} - C\alpha_{i+1}$ and $\vec{r}_{34} = C\alpha_{i+3} - C\alpha_{i+2}$ be the three vectors associated (with abuse of notation, $C\alpha$ denotes here its position).

We define $\vec{i} = \vec{r}_{23} \wedge \vec{r}_{12}$ and $\vec{j} = \vec{r}_{34} \wedge \vec{r}_{23}$ (cf. Figure 8.3). The torsion angle Φ_i is defined as the angle needed to rotate counter clockwise $\vec{i}$ to overlap $\vec{j}$, using $\vec{r}_{23}$ as the up vector.

The basic idea for recovering some torsional information is to analyze the set of known proteins and collect the distribution of angles formed by every sequence of four consecutive aminoacids. In this way, given a sequence (e.g., ALA, GLU, VAL, TYR), we are able to study the statistically preferred torsion angle(s) and thus to model an appropriate energy function, based on a Gaussian approximation of the profile.

In [96] it has been shown that the number of aminoacids influencing a specific torsion angle can be up to 16 neighboring aminoacids. In our case we restrict our attention on a simplified profile. Unfortunately, not enough information is available from the PDB to sharply identify each distribution profile. Even though the PDB contains more than

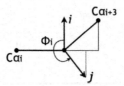

Figure 8.3: Torsion angle Φ_i (projection orthogonal to the axis $C\alpha_{i+1}$–$C\alpha_{i+2}$)

30.000 proteins, roughly 2.000 of them are non redundant (PDB SELECT 25 gives the set of non–homologous proteins, with at most 25% of identities). Thus, even the simplest statistics that considers every combination of types among 4 consecutive aminoacids, requires already a larger set than the entire data base to become meaningful.

Our approach to tackle this lack of information is divided into two parts: first we identify a partition of aminoacids sharing similar torsional properties and afterwards we compute the statistics based on these classes instead of the 20 regular different types of residues. We work with the assumption that if the classes have similar behaviors, the profile for four consecutive residues modulo their class is a good approximation of the distribution and moreover it provides a denser plot. The analysis of data can be split in two parts.

The first analysis aims to the identification of similarities in torsional behaviors among aminoacids. For each aminoacid we collect some specific profile statistics and create a matrix whose entries contain the pairwise Root Mean Square Deviation (RMSD) between the histograms.

In detail, for each aminoacid a, we collect every torsion angle described in the data base, for each of the four positions in which a can appear, namely

$$1 = axyz, \quad 2 = xayz, \quad 3 = xyaz, \quad 4 = xyza$$

where x, y, z ranges in the set of the 20 aminoacids.

We obtain, thus, four histograms a_1, a_2, a_3, and a_4, that form all together the set of data used to compute the similarities. The resulting matrix contains, for each pair of residues, the standard deviation between their two sets of histograms. This information is a good hint in understanding similarities. We treated the standard deviation matrix as a distance matrix and used the algorithm UPGMA (Unweighted Pair Group Method with Arithmetic Mean) as implemented in the program package PHYLIP [87] to generate the tree in Fig. 8.4. We use this tree to cluster sequences.

The clustering output can be used to generate classes according to similarity relation described by the edges. In our case, we decided to employ 4 the classes $\mathcal{G}_1, \ldots, \mathcal{G}_4$ defined in Fig. 8.4.

Using this partitioning, it is possible to compute a distribution profile $\mathcal{P}$ for each 4-tuple $\langle \mathcal{G}_{\pi_1}, \mathcal{G}_{\pi_2}, \mathcal{G}_{\pi_3}, \mathcal{G}_{\pi_4} \rangle$, with $\pi_i \in \{1, \ldots, 4\}$. This statistics is robust, since the number of occurrences in the PDB is much higher than the single aminoacid profiling. Given

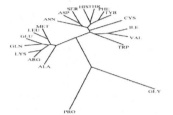

$\mathcal{G}_1$.	ALA ARG GLN GLU LEU LYS MET
$\mathcal{G}_2$.	ASN ASP CYS HIS ILE PHE SER THR TRP TYR VAL
$\mathcal{G}_3$.	PRO
$\mathcal{G}_4$.	GLY

Figure 8.4: The four aminoacid classes equivalent for torsion angles

$C\alpha_i, \ldots, C\alpha_{i+3}$ and their four correspondent aminoacid classes $\mathcal{G}_{\pi_i}, \ldots, \mathcal{G}_{\pi_{i+3}}$, we approximate the corresponding distribution profile with a model made of the sum two Gaussians. In this way, we are able to give an approximation of the torsional distribution for the axis $C\alpha_{i+1}C\alpha_{i+2}$, that usually is composed by *at most two main clear peaks* and some noise. We use the Levenberg-Marquardt method to compute the optimal fitting [193]. In particular, the algorithm computes the six essential parameters to describe the two curves: the amplitudes (a_1, a_2), the variances (σ_1, σ_2) and shifts of the two curves (φ_1, φ_2). Note that the distribution of torsion ranges lies in the interval $[0, 2\pi]$. In the following Figures, we give two examples of distributions we computed: the first is for the 4 class profile associated to ALA ALA ALA ALA (note the sharp peak in correspondence of the typical helix torsion angle) and the second for the profile ALA GLY GLY ALA.

i	1	2
a_i	1.120778	0.032578
φ_i	5.346587	3.800901
σ_i	0.143748	3.319605

i	1	2
a_i	0.514728	0.291777
φ_i	4.216903	1.154707
σ_i	2.048899	1.594432

Figure 8.5: Values for AAAA and AGGA torsional distribution

Figure 8.6: Approximated torsional distribution of AAAA and AGGA classes

In order to obtain energetic terms, we compute the opposite logarithm of the distribution function. The function has the form:

$$E_t(\mathbf{x}, \mathbf{t}) = \sum_{i=1}^{n-3} -\log\left(a_1\, e^{\frac{(\Phi_i - \varphi_1)^2}{(\sigma_1 + \sigma_0)^2}} + a_2\, e^{\frac{(\Phi_i - \varphi_2)^2}{(\sigma_2 + \sigma_0)^2}} \right) \tag{8.8}$$

where the parameters $a_1, a_2, \sigma_1, \sigma_2, \varphi_1, \varphi_2$ depend on $\langle \mathcal{G}_{\pi_i}, \ldots, \mathcal{G}_{\pi_{i+3}} \rangle$ obtained from the vectors $(x_i, x_{i+1}, x_{i+2}, x_{i+3})$, e is the natural base of logarithms and Φ_i is computed as described at the beginning of this section. The parameter σ_0 is used to adapt the distribution variance to an effective energy function, since otherwise we would get too thin holes of potential. In our simulations we set $\sigma_0 = 0.8$.

Note that within this framework we are not taking into account the well known correlation between torsional and bend angles [183].

Contact interaction energy.   The last term of the energy function takes into account contact interactions. For each pair of aminoacids s_i and s_j, such that $|i - j| \geq 3$ we consider a term of the form

$$E_c(\mathbf{x}, \mathbf{t}) = \sum_{i=1}^{n-3} \sum_{j=i+3}^{n} c(x_i, x_j, t_i, t_j) \tag{8.9}$$

$$c(x_i, x_j, t_i, t_j) = |\text{Pot}(t_i, t_j)| \left(\frac{r_0(t_i, t_j)}{r(x_i, x_j)} \right)^{12} + \text{Pot}(t_i, t_j) \left(\frac{r_0(t_i, t_j)}{r(x_i, x_j)} \right)^{6} \tag{8.10}$$

where $r(x_i, x_j)$ is the distance between the $C\alpha$ of x_i and x_j and $r_0(t_i, t_j)$ is a parameter describing the steric hindrance between a pair of non consecutive aminoacids i and j. In our model, $r_0(t_i, t_j)$ is the sum of the radii of the two spheres that represent the aminoacids. The radius depends on the kind of the aminoacid involved. An approximation of them is derived in [90], and the values are reported in Table 8.7. Observe that this value ranges from 3.8 Å (a pair of GLY) to 5.8 Å (a pair of TRP) and it is therefore greater than or equal to the value of 3.8 Å used for two consecutive aminoacids. $\text{Pot}(t_i, t_j)$, instead, is the contact potential between two amino acids, depending on their type. In this energy function, we use the contact matrix developed by Fogolari in [17].

Name	Radius	Name	Radius
GLY	1.90	ALA	1.96
SER	2.02	CYS	2.18
THR	2.21	ASP	2.23
ASN	2.25	PRO	2.28
VAL	2.34	GLU	2.40
GLN	2.41	LEU	2.47
ILE	2.48	HIS	2.49
LYS	2.53	MET	2.54
PHE	2.70	TYR	2.72
ARG	2.73	TRP	2.94

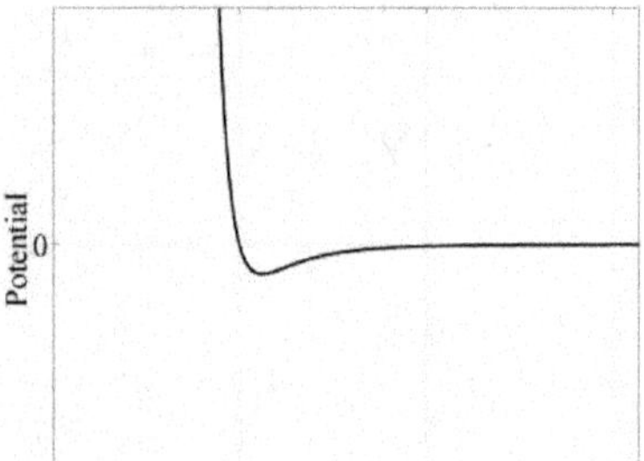

Figure 8.7: Approximated aminoacid radii (in Å) and interaction potential example

Formula (8.10) is similar to typical Lennard-Jones potentials, but differs for those pairs of aminoacids that have unfavorable (i.e. positive) contact energy. The modulus guarantee that short range interactions are repulsive as should be due to steric hindrance. In standard Lennard-Jones potential two atoms always attract at long distance and always repel at short distances. In our recasting of a contact energy in Lennard-Jones form this is true only for those entities having negative contact energy. In this case, the function has a behavior of the kind shown in Figure 8.7. Let us observe that there is a minimum when the distance is 1.12 times greater than $r_0(t_i, t_j)$. Finally note that aminoacids at distance of two in the primary sequence are not considered in E_c, as in [232].

Setting the parameters. A crucial problem while dealing with energy functions is to weight correctly the different terms involved, in order to have the global minimum close to native state of proteins. These parameters have been trained using a sampling of 700 proteins from PDB database with less than 25% homology. Essentially, weights are set in order to minimize the distance between the energy value of the native state and the value of the global optimum. Numerical values can be found in http://www.dimi.uniud.it/dovier/PF/.

8.2 The Simulation Framework

In this section we describe the abstract framework of the simulation, following the line of [30]. This scheme is independent both on the spatial model of the protein and on the energy model employed. Therefore, it can be instantiated using different representations. Here we test the system with the two different energy functions of Section 8.1.

Each amino acid is associated to an independent agent, which moves in the space and communicates with others in order to minimize the energy function. Moreover, we also introduce other agents at different hierarchical levels, which have the objective of coordinating and improving the overall performance of the system.

Milano and Roli in [169] have devised a general scheme to encode agent–based minimization, and our framework can be seen as an instantiation of that model. In particular, they identify four levels of agents, which interact in order to perform the optimization task. Level 0 deals with generation of an initial solution, level 1 is focused on the stochastic search in the state space, level 3 performs global strategic tasks and level 4 is concerned with cooperation strategies.

According to this scheme, here we have agents of level 1, 2 and 3 (level 0 ones are trivial), see Figure 8.8 for a visual representation. We deal separately with them in next subsections.

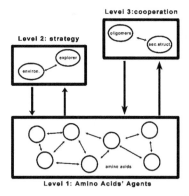

Figure 8.8: Structure of the multi-agent simulation. Black boxes represent the levels, blue circles the agents and red arrows the communications.

We present the basic functions of those agents using sCCP. The code that we will show hereafter is compact, and many implementation details are left apart. This choice is dictated by the fact that our main aim in this Chapter is to discuss the minimization framework, more than the utilization of sCCP to code it. In addition, most of the work consists in routine declarative programming activity in Prolog, thanks to its usage in the definition of constraints (see Section 6.5 of Chapter 6). Probably, the most interesting part of using sCCP resides in the exploitation of its stochastic semantics to encode the stochastic ingredient in the exploration of the search space. However, we deal with these aspects separately in Appendix A. The use of sCCP as model langauge represents a novelty w.r.t. [30, 38], where we used LINDA [52] as concurrent paradigm. Note that some of the variables in this implementation are real-valued, so we need to use constraints over reals rather than on finite domains, as we did in Section 6.5.1 of Chapter 6.

8.2.1   Searching Agents

We associate to each amino acid an agent, which has the capability to communicate its current position to other processes, and to move in the search space, guided by its knowledge of the position of other agents. The general behaviour of these agents can be easily described in sCCP by the process amino(I, N):

$$
\begin{aligned}
&\text{amino}(I, N) :- \\
&\quad \text{ask}_{\lambda_{fast}}(\texttt{authorization}(I)). \\
&\quad \text{tell}_{\infty}(\texttt{remove\_trigger}(I)). \\
&\quad \text{tell}_{\infty}(\texttt{communicate\_moving}(I)). \\
&\quad \exists_{\vec{P}, Newpos}.(\\
&\quad\quad \text{tell}_{\infty}(\texttt{get\_positions}(\vec{P})). \\
&\quad\quad \text{update\_position}(I, N, \vec{P}, Newpos). \\
&\quad\quad \text{tell}_{\infty}(\texttt{new\_position}(I, Newpos)). \\
&\quad).\text{tell}_{\infty}(\texttt{set\_triggers}(I, N)). \\
&\quad \text{tell}_{\infty}(\texttt{remove\_moving}(I)). \\
&\quad \text{amino}(I, N)
\end{aligned}
$$

The input variables of this process are its number I in the chain of aminoacids and the length N of such chain. The first instruction is an ask that checks if the agent is authorized to move. There are two boolean (stream) variables[1] governing this condition. The first one, stored in an auxiliary predicate authorized(I, X), is used to coordinate the activity with higher level agents: these processes can set it to zero, thus blocking the activity of the amino agents. The second, instead, is stored in trigger(I, X), and act as a switch that is set to zero after the move of an agent (with the constraint remove_trigger(I)), and set back to one by another amino agent, after its move. This is a mechanism used to guarantee (a week form of) fairness to the system: each agent must wait for the movement of another process before performing its own move. In this way we avoid that a single agent takes the system resources all for itself. Clearly, at the beginning of the simulation the switches for all the amino processes are turned on, in order to let them move. The asked constraint authorization(I) simply checks if these two variables are equal to one, and it is defined by the following instructions:

```
authorization(I) :-
    authorized(I,X),
```

---

[1]These variables are of type stream, as they change values during the execution of the agent.

```
X #= 1,
trigger(I,Y),
Y #= 1.
```

We are using a notation consistent with Section 6.5 of Chapter 6 and with the constraint logic programming library of SISCtus Prolog [93], where #= denotes an equality constraint. Note that our ask instruction has a rate function associated to it, as sCCP syntax prescribes. The function λ_{fast} is assumed to return a very big constant, so that this instruction gets executed with probability nearly one each time it is encountered.[2]

If this guard is satisfied, the process first sets its trigger switch to zero (telling remove_trigger(I)) and then, telling communicate_moving(I), it sets another boolean switch to one. The corresponding boolean variable (stored in a predicate moving(I,X)) is used in the coordination task, and it allows the amino agent to tell other processes if he is moving or not. In fact, at the end of an execution cycle of amino, it is set again to zero.

After this basic communication activity, the process declares two local variables, $\vec{P}$ and *Newpos*. The first one is indeed a vector of variables that will be used to store temporary the current position of other agents, while the second will store the new position of the agent, after the execution of a move.

The next constraint told, i.e. get_positions($\vec{P}$), retrieves the most recent positions of all other amino acids, which are stored in terms of the kind pos(J,Pos).[3] We omit the presentation of this and future prolog predicates defining constraints, due to their simplicity.

Successively, the current position of each agent is updated by calling the procedure update_position, using a mechanism described in Section 8.2.1. The newly computed position is then posted in the constraint store by telling the new_position(I,Newpos) constraint. Finally, the switches of all other processes are turned on by the constraint set_triggers(I,N) and then the process recursively calls itself.

The position of these agents is expressed in *cartesian coordinates*. This choice implies that the moves performed by these processes are local, i.e. they do not affect the position of other amino acids. This is in syntony with the locality of the potential effects: the modification of the position of an amino acid influence only the nearby ones.

The initial configuration of the chain can be chosen between three different possibilities: straight line, random, and the deposited structure for known proteins.

Simulating moves

The amino acids move according to a simulated annealing scheme. This algorithm, which is inspired by analogy to the physical process of slowly cooling a melted metal to crystallize it, uses a Monte Carlo-like criterion to explore the space, see Chapter 5 for further details. Each time the procedure update_position is invoked, the amino acid I computes a new position p' in a suitable neighborhood (see next Subsection), and then compares its *current* potential P_c with the *new* potential P_n, corresponding to p'. If $P_n < P_c$, the amino acid

---

[2]The mechanism is similar to the one we used in Section 3.2.3 of Chapter 3 for the model of Lotka-Volterra system in restricted PEPA.

[3]Indeed, *Pos* here is a triple of values, as each position of each amino acid is identified by its cartesian coordinates. However, we put off this detail for sake of clarity.

updates its position to p', otherwise it accepts the move with probability $e^{-\frac{P_n - P_c}{Temp}}$. This hill-climbing strategy is performed to escape from a local minimum.

Temp is the parameter controlling the acceptance ratio of moves that increase the energy. In simulated annealing algorithms, it is initially high, and then it is slowly cooled to 0 (note that if $Temp$ is very low, the probability of accepting moves which increase the energy is practically 0). See again Chapter 5 for a more detailed presentation. We discuss the cooling schedule we implement in Section 8.2.2.

The procedure update_position has some random choices in it, and this is where the stochastic semantics of sCCP enters into play. However, the discussion of the encoding of Simulated Annealing in sCCP is rather technical, hence it is deferred to Appendix A.

Moving Strategy

In the energy models adopted (cf. Section 8.1), we have several constraints on the position of the amino acids, which are implemented via energy barriers. This means that we may perform moves in the space that violate these constraints, but they will be less and less probable as the temperature of the system decreases. We have chosen this "soft constrained" approach because the Monte Carlo-like methods require an unbiased exploration of the search space, i.e. they require that the underlying Markov Chain model is irreducible and ergodic (cf. Chapter 5 or [3]). Now, because all the search space is virtually accessible, the moving strategy is very simple, and guarantees the previous two properties: we choose with uniform probability a point in a cube centered at the current position of the amino acid. The length of the side of the cube is set experimentally to 1 Å. Details of the implementation of this choice in sCCP can again be found in Appendix A.

Communication Scheme

When a single agent selects a new position in the space, it calculates the variation of energy corresponding to this spatial shift. However, in almost all the reduced models, the energy relative to a single amino acid depends only from the adjacent amino acids in the polymer chain, and from other amino acids that are close enough to trigger the contact interactions. Thus, if an agent is very far from the current one, it won't bring any contribution to the energy evaluation, at least as long as it remains distant.

This observation suggests a strategy that reduces considerably the communication overhead. Agent i first identifies its spatial neighbors $\mathcal{N}_i$, i.e. all the amino acids in the chain that are at distance less than a certain threshold, fixed here to 14 Å(more or less four times the distance of two consecutive amino acids). Then, for an user-defined number of moves M, it communicates just with the agents in $\mathcal{N}_i$, ignoring all the others. When the specified number M of interactions is reached, amino acid i retrieves the position of all the amino acids and then refresh its neighbour's list $\mathcal{N}_i$.

In our sCCP framework, this means that the process i will turn on the switches trigger(j,X) just for the agents $j \in \mathcal{N}_i$. Moreover, it will read the current positions only of those amino acids, before performing the move.

The refresh frequency must not be too low, otherwise a far amino acid could, in principle, come very close and even collide with the current one, without any awareness of what is happening. We experimentally set it to 100.

This new communication strategy impose a redefinition of the agent behaviour, which now has to update also the neighbor list, leading to the code shown below.

$\text{amino}(I, N, C, NeighList)$:-
 $\text{ask}_{\lambda_{fast}}(\texttt{authorization}(I))$.
 $\text{tell}_\infty(\texttt{remove\_trigger}(I))$.
 $\text{tell}_\infty(\texttt{communicate\_moving}(I))$.
 $\text{update\_neighbors}(C, NeighList, N)$.
 $\exists_{\vec{P}, Newpos} \cdot ($
  $\text{tell}_\infty(\texttt{get\_positions}(\vec{P}, NeighList))$.
  $\text{update\_position}(I, N, \vec{P}, NeighList, Newpos)$.
  $\text{tell}_\infty(\texttt{new\_position}(I, Newpos))$.
 $).\text{tell}_\infty(\texttt{set\_triggers}(I, NeighList))$.
 $\text{tell}_\infty(C\# = C + 1)$.
 $\text{tell}_\infty(\texttt{remove\_moving}(I))$.
 $\text{amino}(I, N, C, NeighList)$

It is quite similar to the one presented in section 8.2.1, with few obvious modifications in the constraint `get_positions` and in the procedure `update_position`, and with the presence of a new procedure for updating the neighbour's list, i.e. `update_neighbors`. This agent performs the update only if C mod M $\equiv 0$, otherwise it leaves $NeighList$ unaltered. Note that both C and $NeighList$ are stream variables.

8.2.2 Strategy

The first layer of agents is designed to explore the search space, using a simulated annealing strategy. However, the neighborhood explored by each agent is small and simple, while the energy landscape is very complex. This means that the simulation is unwilling to produce good solutions in acceptable time periods. This fact points clearly to the need of a more coordinated and efficient search of the solution space, which can be achieved by a global coordination of the agents. This task can be performed by a higher level agent, which has a global knowledge of the current configuration, and it is able to control the activity of the single agents. Details are provided in next subsection.

At the same time, the simulated annealing scheme is based on the gradual lowering of the temperature, which is not a property of the amino agents, but it is rather a feature of the environment where they are endowed. This means that the cooling strategy for temperature must be governed by a higher level agent, which is presented in subsection 8.2.2.

Enhanced exploration of state space

As said in the previous paragraphs, the way the solution space is searched influences very much the performance at finite of stochastic optimization algorithms. Our choice of using an agent based optimization scheme results in an algorithm where subset of variables of the system are updated independently by different processes. However, the neighborhood of each agent is quite restricted, to avoid problems arising from delayed communication. This choice implies that the algorithm may take a very long time to find a good solution, and consequently the temperature must be cooled very slowly.

To improve the efficiency, we designed a higher level agent, called the *"orchestra director"*, which essentially moves the amino acids in the state space according to a different, global strategy. It can move the chain using two different kind of moves: *crankshaft* and *pivot* (cf. Figure 8.9). Crankshaft moves essentially fix two points in the chain (usually at distance 3), and rotate the inner amino acids along the axis identified by these two extremes by a randomly chosen angle. Pivot moves, instead, select a point in the chain (the *pivot*), and rotate a branch of the chain around this hub, again by a random angle. These global moves keep fixed the distance between two consecutive C_α carbon atoms, and are able to overcome the energy barriers introduced by the distance penalty term.

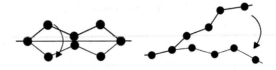

Figure 8.9: Crankshaft move (left) and pivot move (right)

These are essentially the two moves executed by our director, which is described by the following code:

```
director(N, M) :-
        ask_{λfast}(director_authorization).
        tell_∞(communicate_moving_director).
        ∃_{P⃗, P⃗new}.(
        tell_∞(get_positions(N, P⃗)).
        move_chain(M, N, P⃗, P⃗new).
        tell_∞(set_positions(N, P⃗new)).
        ).tell_∞(remove_moving_director).
        director(N, M)
```

The first two lines implement a synchronization mechanism similar to the one used for amino agents. The agents first checks for a boolean variable giving the authorization to move (by asking constraint `director_authorization`), and then it signals its movement by setting an appropriate boolean switch with `communicate_moving_director`. Once the director has got the position of all the amino acids (telling `get_positions`), it performs some moves using the previously described strategies, calling the `move_chain` procedure. This predicate calls itself recursively a predefined number M of times (depending on the protein's length), and each time it selects what move to perform (i.e. crankshaft or pivot), the pivot points and an angle. Then it computes the energy associated to the old and the new configuration, and applies a Monte Carlo criterion to accept the move (cf. Section 8.2.1). Finally, the end of the movement is signaled by constraint `remove_moving_director`.

In the real implementation, we decided to activate this agent only if the temperature of the system is high enough, that is to say, only when this moves guarantee an easy overcome of energy barriers, hence an effective exploration of the search space. Thus we have an additional condition at its beginning, like *Temperature > Threshold*.

Environment

The environmental variables of the simulation are managed by a dedicated agent. In this case, the environment simply controls the temperature, which is a feature of the simulated annealing algorithm. This temperature must not be conceived as a physical quantity, but rather as a control value which governs the acceptance ratio in the choice of moves increasing the energy.

From the theory of simulated annealing (cf. [3] and Chapter 5), we know that the way the temperature is lowered is crucial for the performance of the algorithm. In fact, this can be seen as a sequence of Markov chain processes, every one with its own stationary distribution. These distributions converge in the limit to a distribution which assigns probability one to the points of global minimum for the energy. For this to happen, however, one must lower the temperature logarithmically towards zero (giving rise to an exponential algorithm). Anyway, to reach good approximations, one has just to let the simulation perform a sufficient number of steps at each value of the control parameter, in order to stabilize the corresponding Markovian process.

We applied a simple and common strategy: at each step the temperature is decreased according to $T_{k+1} = \alpha T_k$, where $\alpha = 0.98$. The starting temperature T_0 must allow a high acceptance ratio, usually around 60%, and we experimentally set T_0 as to attain this acceptance ratio at the beginning of the simulation. This value guarantees also that the system does not accept moves with too high energy penalties (cf. Section 8.3). Regarding the number of iterations at each value of T, we set it in such a way that the average number of moves per amino acid is around 500, in order to change the value to all the variables a suitable number of times. The code for the environment agent is straightforward, hence omitted.

8.2.3 Cooperation

In this section we present a dynamic cooperation strategy between agents, which is designed to improve the folding process and try to reach sooner the configuration of minimum energy. The main idea behind is to combine concurrency and some external knowledge to force the agents to assume a particular configuration, which is supposed to be favorable. More specifically, this additional information can be extracted from a database, from statistical observations or from external tools, such as secondary structure predictors.

The cooperation is governed by a high level agent, which has access to the whole status of the simulation and to some suitable external knowledge. The basic additional information we use to induce cooperation is related to the secondary structure. In particular, we try to favor from the beginning the formation of local patterns supposed to appear in the protein (see below).

To coordinate the action of single agents and let a particular configuration emerge from their interaction, we adopt a strategy which is very similar in spirit to *computational fields*, introduced by Mamei [163]. The idea is to create a virtual force field that can drive the movement of the single agents towards the desired configuration. In our setting we deal with energy, not with forces, so we find more convenient to introduce a biasing term modifying the potential energy calculated by a single agent. In this way, we can impose a particular configuration by giving an energy penalty to distant ones (in terms of RMSD).

In addition, the cooperation agent controls the activity of all lower level agents, in particular, it decides the scheduling between the amino agents and the orchestra director. Moreover, it controls also the termination of the simulation. The stopping condition is simple: it is triggered when the system reaches a frozen configuration at zero temperature[4], meaning that a (local) minimum has been reached and no further improvement is possible. The sCCP code of the agent is the following:

$$\text{cooperator}(Moves, \vec{P}_{old}, SecInfo, N) :-$$
$$\exists_{\vec{P}_{new}}($$
$$\text{tell}_\infty(\texttt{get\_positions}(N, \vec{P}_{new})).$$
$$\text{tell}_\infty(\texttt{secondary\_cooperation}(\vec{P}_{new}, SecInfo)).$$
$$\text{check\_termination}(\vec{P}_{old}, \vec{P}_{new}).$$
$$\text{tell}_\infty(\texttt{give\_authorization\_amino}(N)).$$
$$\text{ask}_{\lambda_{fast}}(\texttt{amino\_moves}(Moves)).$$
$$\text{tell}_\infty(\texttt{remove\_authorization\_amino}(N)).$$
$$\text{ask}_{\lambda_{fast}}(\texttt{amino\_resting}(N)).$$

---

[4]When the temperature falls below a predefined threshold, it is quenched to zero.

$\text{tell}_\infty(\texttt{give\_authorization\_director}).$
$\text{ask}_{\lambda_{delay}}(true).$
$\text{tell}_\infty(\texttt{remove\_authorization\_director}).$
$\text{ask}_{\lambda_{fast}}(\texttt{director\_resting}).$
$\text{cooperator}(Moves, \vec{P}_{new}, SecInfo, N)$)

The first actions of the agent are reading the current configuration of the protein (telling constraint **get_positions**), activating the secondary structure cooperation mechanism (telling constraint **secondary_cooperation**) and checking the termination conditions (using the procedure **check_termination**). In particular, the activation of the cooperation mechanism is obtained by switching on some boolean variables that trigger the corresponding energy term in the computations performed by lower level agents. The following predicates are devoted to the synchronization between the amino agents and the orchestra director. The amino agents are authorized to move by telling the constraint **give_authorization_amino**. When the cooperator decides that amino agents should stop, it tells constraint **remove_authorization_amino** and waits for the agents to stop moving (by asking **amino_resting**). Things work similarly for the orchestra director. The asked constraint **amino_moves** becomes entailed when the amino acids have performed a certain amount of moves, specified in the variable $Moves$. Note that we introduce a stochastic delay in between giving and removing authorization to the director, in order to let it start moving (with high probability, as λ_{delay} will be much smaller than λ_{fast}).

In the previous code, we showed a strategy that alternates between amino agents and orchestra director, by a mutually exclusive activation. Another possible strategy is to run both amino agents and the orchestra director simultaneously, though this may introduce more errors in the simulated annealing scheme due to higher asynchrony. A comparison of the two strategies can be found in Section 8.3.

Cooperation via Secondary Structure

In the literature it is recognized that the formation of local patterns, like α-helices and β-sheets, is one of the most important aspects of the folding process (cf. [179] and Chapter 4). Actually the combined usage of alignment profiles, statistical machine learning algorithms and consensus methods has resulted in location of these local structures with a three state accuracy close to 80% (see for a discussion [208] and references cited therein). We plan to use the information extracted from them to enhance the simulation. For the moment, however, we introduced a preliminary version of cooperation via secondary structure, which identifies the location of secondary structure directly from pdb files (when using information coming from secondary structure predictors, we should include in the cofields also the accuracy level of the prediction).

Once the cooperation agent possesses this knowledge, it activates a computational field that forces amino acids to adopt the corresponding local structure. The mathematical form of this new potential penalizes all configurations having a high RMSD from a "typical" helix or sheet.[5] Of course, this energy regards only the amino acids supposed to form a secondary structure, and it is activated from the beginning of the simulation, as it should be able to drive the folding process, at least locally.

8.2.4   Implementation

The code presented in the previous section can be executed with the interpreter that has been designed for sCCP, see Chapter 6. However, the number of moves that the system has to perform in order to produce decent predictions is very high, thus these computational issues

---

[5]There are different kinds of α-helices and β-sheets, but we omit here further details.

First Energy Model	Energy	RMSD
Without orchestra director	-6.688 (1.483)	25.174 (0.863)
With orchestra director	-56.685 (2.518)	7.644 (0.951)

Second Energy Model	Energy	RMSD
Without orchestra director	28.944 (5.500)	12.892 (0.825)
With orchestra director	11.325 (3.882)	8.404 (1.078)

Table 8.1: Comparison of performances of the framework with and without the introduction of the orchestra director agent, for the first energy model (top table) and the second one (bottom table). Values are averages over ten runs, variance is shown in brackets.

cannot be neglected. To tackle this problem, we have also written a multithreading version in pure **C**, which can run both under Windows. This version reproduces the communication mechanisms of the sCCP code, using the shared memory, so it is equivalent to the program presented in the paper, though much more efficient. All the codes can be found in `http://www.dimi.uniud.it/dovier/PF`. Note that all experimental tests have been performed using this implementation.

8.3 Experimental results

In this section we present the results of some tests of our program. We are mainly interested in two different aspects: seeing how the features of the framework (parallelism, strategy, cooperation) influence the simulation and checking how good are the predicted structures with respect to the resolution of the energy functions used. Note, however, that these potentials are structurally very simple, so we are not expecting outstanding results out of them.

We ran the simulation on different proteins of quite small size, taken from PDB [16]. This choice is forced by the low resolution of the potential. We also had to tune a lot of parameters of the program, especially the cooling schedule for the temperature, the weights of the penalty terms and the scheduling of the strategic and the cooperative agents.

All the tests were performed on a bi-processor machine, mounting two Opteron dual core CPU at 2 GHz. Therefore, we have a small degree of effective parallelism, which can give a hint on the parallel performances of the system.

In the following, we comment separately on the agent-based framework and on the energy models.

8.3.1 Analysis of the Agent-Based Framework

In order to analyze the different features of the system, we performed several tests on a single protein, 1VII, which is composed of 36 amino acids. In the following we focus our attention on the orchestra director, on the cooperation features and on the effects of parallelism on the underlying simulated annealing engine.

Exploration Strategy To evaluate the enhancements introduced by the orchestra director, we ran several simulation with and without it, comparing the results, in terms of energetic values and RMSD from the native state of the solutions found. As expected, the amino acid agents alone are not able to explore exhaustively the state space, and the minimum values found in this case are very poor (see Table 8.1). This depends essentially from the fact that most of their

First Energy Model	Energy	RMSD	time (min)
Sequential	-54.068 (1.949)	10.728 (0.793)	280
Multi-Agent	-56.685 (2.518)	7.644 (0.951)	85

Second Energy Model	Energy	RMSD	time (min)
Sequential	39.599 (5.072)	11.255 (0.837)	28
Multi-Agent	11.325 (3.882)	8.404 (1.078)	10

Table 8.2: Comparison of the multi agent scheme with a sequential simulated annealing performing pivot and crankshaft moves. Top table compares results for the first energy model (for protein 1VII), while bottom table deals with the second energy model. Both simulations execute the same number of moves per amino acid. Multi-agent simulation performs better in all aspects, and it is notably three times quicker, using 4 processors. Values are averages over ten runs, variance is shown in brackets.

moves violate the distance constraint, especially at high temperatures. At low temperatures, instead, the system seems driven more by the task of minimizing this penalty term, than by the optimization of the "real" components of energy. Therefore, the simulation gets stuck very easily in bad local minima, and reaches good solutions just by chance. If the "orchestra director" agent is active, instead, much better energy minima are obtained (see again Table 8.1). Note that combining the amino acid agents and the strategic one corresponds to having a mixed strategy for exploring the state space, where two different neighborhoods are used: the first one, local and compact, is searched by the amino acid agents, while the second one, which links configurations quite far away, is searched by the strategic agent.

Figure 8.10: Two structures for 1VII generated using the second energy model, with (right) and without (left) orchestra director.

We observe also that the poor values of the energy and of RMSD for the simulation with amino agents alone depend from the fact that this exploration scheme is not able to compact and close the structure by itself. In fact, the chain remains open, and we can see only some local structure emerging. This is evident in Figure 8.10.

Effects of Parallelism It is well-known that asynchronous parallel forms of simulated annealing can suffer from a deterioration of results with respect to sequential versions (cf. [112]), due to the use of outdated information in the calculation of the potential. Therefore, we ran some sequential simulations, using essentially the moves performed by the orchestra director, that is crankshaft and pivot. The energy values of solutions obtained with the sequential simulation are, on average, a little bit higher than those of the multi-agent simulation (see Table 8.2).

Protein	# amino	helix	sheet
1LE0	12		x
1KVG	12		x
1LE3	16		x
1EDP	17	x	
1PG1	18		x
1ZDD	34	x	
1VII	36	x	
2GP8	40	x	
1ED0	46	x	x

Table 8.3: List of tested proteins, with number of amino acids and type of secondary structure present.

Hence, the introduction of parallelism not only does not worsen the quality of solutions, but it also slightly improves it. This may depend on the fact that amino agents are able to locally intensify the exploration of the search space. In addition, we compared the execution time of the parallel and the sequential simulations, for the same total number of moves per amino acid. As we can see, multi-agent simulation is three times quicker using 4 processors, showing therefore an almost optimal parallel speed-up.

Cooperation To test the cooperative field, we compared separately runs with and without cooperation via secondary structure. It comes out that the quality of the solutions in terms of energy are generally worse with the cooperative agent active (cf. Tables 8.4 and 8.5, and 8.6 and 8.7). On the other hand, the RMSD is improved in most of the cases (cf. again Tables 8.4 and 8.5, and 8.6 and 8.7). This is quite remarkable, as the information relative to secondary structure is still local. We are also pondering the introduction of a better form of cooperation, with some capability of driving globally the folding process, cf. Section 8.3.3 for further comments.

8.3.2 Energy Comparison

In this section we compare the performances of the two energy models presented in Section 8.1. The quantities used to estimate the goodness of the results are the value of the energy and the root mean square deviation (RMSD) from the known native structure.

In the following, we presents the results of tests ran on a set of 9 proteins, extracted from the Protein Data Bank. The complete list of tested proteins, together with their length and the type of secondary structure present, is shown in Table 8.3. Therefore, knowing their native structure, we are able to estimate the distance of the predicted structure from the real one, thus assessing the quality of the potentials. In the tables following, we present the simulation results with and without cooperation.

First Energy Model In Table 8.4, we show the best results obtained in terms of RMSD and energy without cooperation, while in Table 8.5 cooperation was active. For a safe comparison, the energy value does not take into account the contributions of cooperative computational fields, present in experiments of Table 8.5 only. The numbers shown are an average of 10 runs; standard deviation is shown in brackets.

Protein	Energy	RMSD
1LE0	-11.5984 (0.8323)	4.0324 (0.7783)
1KVG	-19.8048 (0.7838)	3.8898 (0.6206)
1LE3	-29.5412 (1.1330)	6.0749 (0.7525)
1EDP	-62.9978 (1.5487)	5.0703 (0.7261)
1PG1	-52.7987 (1.5356)	7.1490 (0.8791)
1ZDD	-45.4133 (1.9326)	8.4255 (1.2578)
1VII	-56.6850 (2.5189)	7.6447 (0.9509)
2GP8	-10.5285 (1.7073)	8.7200 (1.1488)
1ED0	-50.8481 (2.5158)	9.3103 (1.1701)

Table 8.4: Results for the first energy model without cooperation. Values are averages over ten runs, variance is shown in brackets.

Results for the first potential, with cooperation

Protein	Energy	RMSD
1LE0	-10.5248 (0.9756)	4.2291 (0.7025)
1KVG	-16.7263 (0.9267)	3.8555 (0.5476)
1LE3	-27.1925 (0.9693)	5.5847 (0.5870)
1EDP	-49.0921 (1.6160)	3.7743 (0.5743)
1PG1	-33.7463 (1.3319)	4.3540 (0.6856)
1ZDD	-49.6080 (2.2137)	8.2052 (1.1388)
1VII	-56.2129 (1.5908)	7.7737 (0.8158)
2GP8	-2.6426 (1.8351)	4.1871 (0.9585)
1ED0	-57.4477 (2.2198)	8.8542 (1.0181)

Table 8.5: Results for the first energy model with cooperation. Values are averages over ten runs, variance is shown in brackets.

From Table 8.4, we can see that, without cooperation, the simulation is quite stable: most of the runs produce solutions with energy varying in a very small range of values. On the contrary, the RMSD is quite high. This depends mostly on the low resolution of the potential. In fact, this energy function has terms which compact the chain, but no term imposing a good local structure. Therefore, the simulation maximizes the number of contacts between amino acids, giving rise to a heavily non-physical shape. The situation, however, changes with the introduction of the cooperative effects. In this case, the maximization of contacts goes together with potential terms imposing good local shapes, creating better structures but worsening the energy (less contacts are formed). In particular, we can see a remarkable improvement of RMSD for proteins 1PG1 and 2GP8; a visual comparison of the outcome for 2GP8 is shown in Figure 8.11.

On the other hand, for several proteins, among which 1VII, cooperation does not bring any sensible improvement on the RMSD value. This probably depends from the fact that, in these cases, the constraints on the secondary structure are not so strong to force a good global shape. In particular, the predominant terms of the energy function try to maximize the number of favorable contacts, thus creating a structure where some areas are extremely compact and others are left open. If the local effect induced by secondary structure cofields is concentrated in the compressed area, then no essential improvement arises. This is a witness of the low

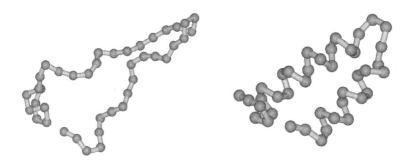

Figure 8.11: Protein 2GP8 predicted with the first potential. From left to right: a solution without cooperation and a solution with cooperation. Native state is depicted in Figure 8.12.

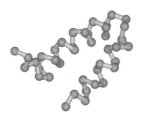

Figure 8.12: Native state for 2GP8.

resolution of the potential.

Second Energy model This energy model was never tested in all its potentiality before. Test in [28] were performed in a simulation with only amino agents, which are not able to overcome the energy barriers blocking the compactification of the chain.

In Tables 8.6 and 8.7, we can see the results for the set of testing proteins listed in Table 8.3. We can see that also the resolution of this potential is not very accurate. Generally, the values in terms of RMSD are slightly worse than for the first energy model. This happens despite the presence of stronger local terms, that should, at least in principle, generate better shapes. Specifically, those terms seem to enter in conflict with the contact energy, forcing a shape with good local properties but poor global structure. For instance, for 1ZDD without cooperation, the chain remains open, though some form of helices emerge, cf. Figure 8.13.

Secondary structure cooperation follow a pattern similar to the first energy model, improving the RMSD (and worsening the energy) for a slightly broader set of proteins. For example, 1ZDD gets closed with secondary structure cooperation, see Figure 8.13.

Comparing the first and the second energy model, we can observe that there is no real winner. The first potential has generally solutions of slightly better quality, but it is much heavier from

Results for the second potential, without cooperation

Protein	Energy	RMSD
1LE0	2.8798 (1.4315)	5.4120 (0.6963)
1KVG	-1.3340 (1.1783)	5.0596 (0.9254)
1LE3	1.8239 (1.7636)	7.2068 (0.9228)
1EDP	-3.5488 (0.7719)	6.1642 (0.7851)
1PG1	23.7913 (1.3656)	9.1229 (1.3274)
1ZDD	14.1963 (3.1172)	8.5113 (1.2836)
1VII	11.3256 (3.8826)	8.4040 (1.0787)
2GP8	48.6291 (4.5198)	7.9520 (1.5060)
1ED0	16.7202 (2.1936)	10.7979 (0.8692)

Table 8.6: Results for the second energy model without cooperation. Values are averages over ten runs, variance is shown in brackets.

Results for the second potential, with cooperation

Protein	Energy	RMSD
1LE0	4.8766 (2.7584)	5.2671 (0.8173)
1KVG	1.7643 (1.4065)	4.9455 (0.8481)
1LE3	1.8885 (1.6943)	6.6555 (1.1844)
1EDP	-4.7164 (0.8152)	6.0124 (0.4821)
1PG1	39.3512 (2.5879)	12.9420 (1.0104)
1ZDD	-10.7000 (0.8261)	6.8331 (1.2393)
1VII	-9.0183 (1.2882)	8.0266 (1.1269)
2GP8	-7.7530 (1.0468)	5.3764 (1.4519)
1ED0	-9.6144 (1.4974)	8.4351 (0.8369)

Table 8.7: Results for the second energy model with cooperation. Values are averages over ten runs, variance is shown in brackets.

a computational point of view (the simulation using the second potential is approximately 10 times faster, cf. Table 8.2). This increase in time depends from the fact that the first model needs continuously the computation positions of C_β atoms. Moreover, also the single terms in the energy are more expensive to calculate.

The previous discussion implies that we need a more refined potential, taking into account all the correlations involved and modeling not only the C_α atoms, but also a coarse-grained representation of the whole side chain.

8.3.3   Conclusions

The agent-based framework presented here shows interesting potentials, though the energy functions used are too coarse to compete with up-to-date ab-initio predictors (see Section 4.4.3 of Chapter 4). In particular, it's remarkable the gain in speed and in quality of solutions with respect to a sequential version of the algorithm (cf. Section 8.3). Currently, we are trying to overcome the problems connected with the low resolution of potentials used, by developing a more reliable energy [91] that is expected to provide much more accurate results.

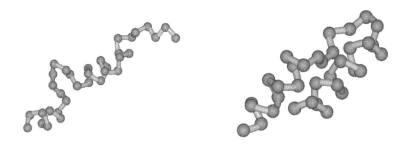

Figure 8.13: Protein 1ZDD predicted with the second potential. From left to right: a solution without cooperation and a solution with cooperation. Native state is depicted in Figure 8.14.

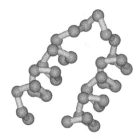

Figure 8.14: Native state for 1ZDD.

The cooperation level is a powerful feature of the framework, though it is not exploited yet. Specifically, it offers the possibility of designing complex heuristics depending on external information and on the search history [169]. Up to now, we use only local information related to secondary structure. One possible direction to improve it is to introduce information about the real physical dynamic of the folding, adding, for instance, a computational field mimicking the hydrophobic force, therefore compactifying the structure at the beginning of the simulation. Another possibility is to identify the so called *folding core* [42], which is a set of key contacts between some amino acids. These contacts have a stabilizing effect, and their formation is thought to be one of the most important steps in the folding process. These contacts can be forced during the simulation by means of a computational field. In general, to fully exploit the cooperativity of agents, we need to integrate this information with exploration-dependant knowledge.

Finally, we plan to integrate this simulation engine in a global schema for protein structure prediction, combining together multi-agent simulation (using a potential modeling also the side-chain), lattice minimization [63] and molecular dynamics.

Chapter 9

Relating sCCP and Differential Equations

In this chapter we tackle the intriguing and difficult problem of relating discrete and stochastic models one one side and continuous and deterministic models on the other. Biochemical reactions can be given two different kinetics, either deterministic or stochastic, see Sections 3.1.3 and 3.1.3 of Chapter 3 for further details. In this context, the question of how these two formalism relate together has been already studied, leading to the development of an hybrid approach via stochastic differential equations to chemical kinetics, i.e. the Chemical Langevin Equation, see Section 3.3.4 of Chapter 3.

Ideally, one would like to have a modeling technique that collects the advantages both of stochastic process algebras and differential equations, or, at least, to switch automatically between the two formalisms, depending on the particular task to be performed. In this direction, there are two related problems that must be faced: (a) studying the (mathematical) relation between the two modeling techniques and (b) finding automatic methods for converting one formalism into the other.

More specifically, we suggest the following workflow: first defining translation methods (for a specific process algebra), thus tackling (b), and then studying the mathematical relations intervening between the models obtained applying these translations. In this way we should be able to evaluate the appropriateness of conversion procedures between SPA and ODE's and to restrict the focus of the analysis required by (a).

There are two directions in the conversion between SPA and ODE: the first one associating a set of differential equations to a stochastic process algebra model, and the inverse one, mapping differential equations to stochastic process algebra programs. The first direction can be helpful for the analysis of SPA models, as ODE's can be solved and analyzed more efficiently. Associating SPA to ODE, instead, can help to clarify the logical pattern of interactions that are hidden in the mathematical structure of differential equations. Generally, as process algebra models can be written much more easily than differential equations, even by non-experts (possibly via a graphical interface), the first direction, from SPA to ODE, looks potentially more fruitful, though having both mappings helps the study of the relationship between the two formalisms.

Supposing to have such transformations at our disposal, a crucial problem is to single out criteria to evaluate and validate them. The first possibility is to inspect the relationship intervening between a SPA program and the associated ODE's only from a mathematical point of view, forgetting any information about the system modeled. As both stochastic processes and differential equations are dynamical systems, this approach essentially corresponds to require that both models exhibit the same behavior, i.e. the same dynamical evolution. Of course, we may require agreement only from a qualitative point of view (so that the qualitative features of the dynamics are the same) of even from a quantitative one (numerical values agree). The difficulty with this approach is that stochastic processes have a noisy evolution, in contrast with the determinism characterizing differential equations. Hence, we need to remove the noise. One possibility is to look only at qualitative features of the dynamics, defining them in a precise way; we will go back to this problem in Section 9.2.3 below. Otherwise, we may average out noise from the stochastic models, thus considering the expected evolution of the system and requiring it to be described precisely (i.e. quantitatively) by the ODE's. Unfortunately, noise cannot be eliminated so easily, as sometimes it is the driving force of the dynamics [102, 234]. Therefore, this second form of equivalence is not completely justified; we will comment more on this point while discussing some examples in the following.

A different approach in comparing stochastic and differential models can be defined if we consider some additional information, which is external to the mathematics of the two models. The idea is to validate the translation w.r.t. this additional information. We explain this point with an example. Consider a model of a set of biochemical reactions; there are different chemical kinetic theories that can be used to describe such system, the most famous one being the principle of mass action. Using such a kinetic theory, we can build (in a canonical way) both a model based on differential equations and a model based on stochastic process algebras. If we are concerned with the principle of mass action more than with dynamical behavior, we may ask that our translation procedures preserves the former, meaning that the ODE's associated to a mass action SPA program are exactly the ODE's built according to mass action principle, and viceversa. Essentially, this corresponds to requiring that the translation procedures defined are *coherent with (some) principles of the system* modeled. For instance, in the case of mass action, coherency corresponds to preserve the meaning of rates (the so called *rate semantics* in [49]). Notice that in this case we are not requiring anything about dynamics, so coherent models may exhibit a divergent behavior, and this is indeed a well known issue, see, for instance, [102] or Section 9.2.3 below. Therefore, this comparison is essentially different from the behavioral-based one, and it is essentially syntactic, in the sense that it is concerned only with how models are written, not with their time evolution.

The operation of associating ODE's to SPA can be seen also as the definition of an ODE-based semantic for the stochastic processes, as opposed to the CTMC-based one. Consequently, the comparison of the stochastic model with the derived ODE's can also be seen as an attempt to discover the mathematical relationship between these two semantics.

The problem of associating ODE's to stochastic process algebras has been tackled only recently in literature. The forefather is the work of Hillston [129], associating ODE's to models written in PEPA [128], a stochastic process algebra originally designed for performance modeling. Successively, similar methods have been developed for stochastic π-calculus [48, 32, 194] and for stochastic Concurrent Constraint Programming [25, 33]. All these methods build the ODE's performing a syntactic inspection and manipulation of the set of agents defining the SPA model. In fact, they all satisfy the coherency condition staten above, at least for mass action principle (a proof for stochastic π-calculus ca be found in [49]). The inverse problem of associating SPA models to ODE's has received much less attention, the only example being [33], where we use stochastic Concurrent Constraint Programming as target SPA.

In this chapter we extend the work done in [32], presenting it in a more detailed and formal way. Basically, we will define two translation procedures: from sCCP to differential equations and viceversa. sCCP plays here a central role, thanks to some ingredients giving a noteworthy flexibility to it, the presence of functional rates above all. The translation procedure will be defined for a restricted version of sCCP, introduced in Section 9.1. The mapping from sCCP to differential equations is presented formally in Section 9.2, while Section 9.3 is devoted to the presentation of the inverse mapping from general ODE's into sCCP. In Section 9.2.2 we show how the ODE's obtained are a first-order approximation of the equation for the time evolution of the average of the CTMC associated to the sCCP program. This argument, however, works only for *synchronization-free* agents, i.e. basically agents without guards. In Section 9.2.1, we will also show coherency conditions for a class of chemical kinetics, while in Section 9.2.3 we will comment in detail the problem of behavioral equivalence in the conversion from sCCP to ODE's. This will be done mainly via examples, exhibiting biological systems for which the translation preservers also the behavior and other systems whose stochastic models show a different behavior than ODE's. The problem of behavioral equivalence is not new, and in fact some examples that we will give are famous ones [102]. However, the syntactic structure of process algebras in general, and sCCP specifically, give a new flavor to these classical examples, and brings the attention into new ones.

The issue of preservation of dynamic behavior in the mapping from ODE's to sCCP is tackled in Section 9.3. In this case we are able to exploit the structure of the mapping and thus to give a convergence theorem.

9.1 Restricted sCCP

The mapping between sCCP and ODE's is not defined for the whole sCCP language, but rather for a restricted version of it, which is, however, sufficient to describe biochemical reaction and genetic networks.

This restricted version of sCCP will be denoted in the following by RESTRICTED($sCCP$), and is formally specified by the following definition:

Definition 13. A RESTRICTED($sCCP$) program is a tuple $(Prog, \mathbf{X}, init(\mathbf{X}))$ satisfying:

1. *Prog* is an sCCP-program respecting the grammar defined in Table 9.1.

2. The variables used in the definition of agents are taken from a finite set $\vec{X} = \{X_1, \ldots, X_n\}$ of *global stream-variables*, each with the same domain $\mathbb{D}$, usually $\mathbb{D} = \mathbb{N}$ or, more generally, $\mathbb{D} = \mathbb{Z}$.

3. The only admissible updates for variables $\{X_1, \ldots, X_n\}$ are constraints of the form $X_i = X_i + k$ or $X_i = X_i - k$, with $k \in \mathbb{D}$ constant.

4. Constraints that can be checked by ask instructions are finite conjunctions of linear equalities and inequalities.

5. The initial configuration of the store is specified by the formula $init(\mathbf{X})$, consisting in the following conjunction of constraints: $(X_1 = x_1^0) \wedge \ldots \wedge (X_n = x_n^0)$, with the constants $x_i^0 \in \mathbb{D}$ referred to as the *initial values* of the sCCP-program.

This definition can be justified looking at the sCCP-agent associated to a biochemical reaction and also at the sCCP-model of genes considered in Chapter 7.

In fact, in these cases all employed variables are numerical variables of the stream-type[1], while all updates in the store add or subtract them a predefined constant quantity. Guards, instead, usually check if some molecules are present in the system ($X > 0$), though we consider here the more general case of linear equalities and inequalities. The use of global variables only, instead, can be justified noting that the existential operator $\exists_x$ is never used (neither in the reaction agent nor in gene models of Chapter 7), as the scope of molecular interactions is system-wide. The suppression of the operator $\exists_x$, as a side consequence, guarantees that we can avoid to pass parameters to procedure calls: in fact, each procedure can be defined as operating on a

---

[1] We do not need further types of variables, as we just need to count the number of different molecules in the system.

specific subset of global variables. However, parameter passing is used in Section 9.2.1 to define parametrically a generalized reaction agent. Therefore, we agree that each instance of a reaction agent, say f-reaction$(\mathbf{R}, \mathbf{X}, \mathbf{P}, \mathbf{k})$, is replaced with the corresponding ground form f-reaction$_{(\mathbf{R}, \mathbf{X}, \mathbf{P}, \mathbf{k})}$. The same trick will be used for other agents. We demand further comments on the restrictions in Section 9.2.5.

In order to fix the notation in the rest of the paper, we give the following definition:

Definition 14. A RESTRICTED($sCCP$) agent A not containing any occurrence of the parallel operator $\|$ is called a *sequential component* or a *sequential agent*.
A RESTRICTED($sCCP$) agent N is called an *sCCP-network* if it is the parallel composition of sequential agents.

Inspecting the grammar of Table 9.1, we can observe that the initial configuration of a RESTRICTED($sCCP$) program is indeed an sCCP-network. The following property is straightforward:

Lemma 7. *The number of sequential components forming an sCCP-program* $(Prog, \mathbf{X}, init(\mathbf{X}))$ *remains constant at run-time and equals the number of sequential agents in the sCCP-network of the initial configuration.*

Proof. As sequential components do not contain any parallel operator, no new agents can be forked at run-time.[2]   ∎

In the rest of the paper, for notational convenience, we usually identify an sCCP-program with the corresponding sCCP-network.
Moreover, forbidding the definition of local variables implies the following property:

Lemma 8. *The number of variables involved in the evolution of an sCCP-network[3] is a subset of* $\{X_1, \ldots, X_n\}$*, hence finite.*

The restrictions of RESTRICTED($sCCP$) are in the spirit of those introduced in [129]: we are forbidding an infinite unfolding of agents and we are considering global interactions only, forcing the speed of each action to depend on the whole state of the system. Indeed, also in [48] we find similar restrictions, though the comparison with sCCP is subtler. First of all, the version of π-calculus presented in [48] does not allow the use of the restriction operator, meaning that interactions have a global scope. However, agents in the π-calculus of [48] are not sequential, as each process is associated to a single molecule and the production of new molecules is essentially achieved by forking processes at run-time. This is not necessary in sCCP, as sCCP-agents model reactions, while molecules are identified by variables of the system. What is finite in [48], however, is the number of syntactically different agents that can be present in a system.

9.2   sCCP to ODE

In this section we define a translation method associating a set of ordinary differential equations to an sCCP program. This translation applies precisely to RESTRICTED($sCCP$) defined above, similar in many ways to the one defined to model check sCCP with PRISM, see Section 7.5 of Chapter 7. The procedure is organized in several simple steps, illustrated in the following paragraphs. Essentially, we first associate a finite graph to each sequential component of an

---

[2]We are counting also deadlocked agents.
[3]A variable is involved in the evolution of the network if one of the following things happen: it is updated in a tell instruction, it is part of a guard checked in an ask instruction, or it is used in the definition of a rate function.

$$\boxed{\begin{array}{ll}
Prog = Def.N & Def = \varepsilon \mid Def.Def \mid p : -A \\
\pi = \text{tell}_\lambda(c) \mid \text{ask}_\lambda(c) & M = \pi.G \mid M + M \\
G = \text{tell}_\infty(c).G \mid p \mid M & A = \mathbf{0} \mid M \\
N = A \mid A \parallel N &
\end{array}}$$

Table 9.1: Syntax of the restricted version of sCCP.

sCCP network and then, analyzing the graph, we define an interaction matrix similar to the one defined in [129] or to action matrices of (stochastic) Petri nets (see, for instance, [117]). Writing ODE's from this matrix is then almost straightforward.

After defining this translation, in Section 9.2.1 we investigate how it relates to biochemical kinetics and we show that the ODE's associated to an sCCP-model of a set of biochemical reactions are the ones generally considered in standard biochemical praxis [49, 58]. Some considerations on dynamical properties are then put forward in Section 9.2.3, while in Section 9.2.4 the focus is moved on the concept of behavioral equivalence. Finally, in Section 9.2.5 we reconsider the restrictions applied to sCCP in the light of the described transformation procedure.

Step 1: Reduced Transition Systems.

The first step consists in associating a labeled graph, called *reduced transition system* [26], to each sequential agent composing the network. As a working example, we consider the following simple sCCP agent:

$$\begin{aligned}
\text{RW}_X :- \\
&\quad \text{ask}_1(X > 0).\text{tell}_\infty(X' = X - 1).\text{RW}_X \\
&+ \text{tell}_1(X' = X + 2).\text{RW}_X) \\
&+ \text{ask}_{f(X)}(true).(\quad \text{ask}_1(X > 1).\text{tell}_\infty(X' = X - 2).\text{RW}_X \\
&\qquad\qquad\qquad + \text{tell}_1(X' = X + 1).\text{RW}_X)
\end{aligned}$$

$$f(X) = \tfrac{1}{X^2+1}$$

This agent performs a sort of random walk in one variable, increasing or decreasing its value by 1 or 2 units, depending on its inner state.

Inspecting Table 9.1, where the syntax of RESTRICTED($sCCP$) is summarized, we observe that each branch of a stochastic choice starts with a stochastic timed instruction, i.e. an $\text{ask}_\lambda(c)$ or a $\text{tell}_\lambda(c)$, followed by zero or more $\text{tell}_\infty(c)$, followed by a procedure call or by another stochastic choice. The first operation that we need to perform, in order to simplify the structure of agents, is that of collapsing each timed instruction with all the instantaneous tell instructions following it and replacing everything with one "action" of the form

$$\text{action}(c, d, \lambda),$$

where c is a guard that must be entailed by the store for the branch to be entered, d is the constraint that will be posted to the store, and λ is the stochastic rate of the branch, i.e. a function $\lambda : \mathcal{C} \to \mathbb{R}^+ \cup \{\infty\}$. The presence of ∞ among possible values of λ is needed to simplify the treatment of instantaneous tells.

$$
\begin{array}{c}
Prog = \hat{Def}.\hat{A} \\
\hat{Def} = \varepsilon \mid \hat{Def}.\hat{Def} \mid [\![p]\!] : -\hat{A} \\
\hat{\pi} = \text{action}(g, c, \lambda) \qquad \hat{M} = \hat{\pi};\hat{G} \mid \hat{M} \oplus \hat{M} \\
\hat{G} = [\![p]\!] \mid \hat{M} \qquad \hat{A} = [\![\mathbf{0}]\!] \mid \hat{M} \\
\hat{N} = \hat{A} \mid \hat{A} \parallel \hat{N}
\end{array}
$$

Table 9.2: Syntax of COMPACT($sCCP$).

To achieve this goal, we formally proceed by defining a conversion function, named COMPACT, by structural induction on terms. The result of this function is that of transforming an agent written in RESTRICTED($sCCP$) into an agent of a simpler language, called COMPACT($sCCP$), where **ask** and **tell** are replaced by the instruction **action**. In order to distinguish between the two languages, we denote stochastic summation in COMPACT($sCCP$) by "$\oplus$", sequential composition by ";", and we surround procedure calls and nil agent occurrences by double square brackets "$[\![\cdot]\!]$". The syntax of COMPACT($sCCP$) is formally defined in Table 9.2; its constraint store, instead, follows the same prescriptions of Definition 13.

In defining the function COMPACT, we use a concatenation operator $\bowtie$ to merge instantaneous tells with the preceding stochastic action. Formally, COMPACT is defined as follows:

Definition 15. The function COMPACT: RESTRICTED(sCCP) $\rightarrow$ COMPACT(sCCP) is defined by structural induction through the following rules:

1. COMPACT($\mathbf{0}$) = $[\![\mathbf{0}]\!]$;

2. COMPACT(p) = $[\![p]\!]$.

3. COMPACT($\text{ask}_\lambda(c).G$) = $\text{action}(c, true, \lambda)$ $\bowtie$COMPACT(G).

4. COMPACT($\text{tell}_\lambda(c).G$) = $\text{action}(true, c, \lambda)$ $\bowtie$COMPACT(G).

5. COMPACT($\text{tell}_\infty(c).G$) = $\text{action}(true, c, \infty)$ $\bowtie$COMPACT(G).

6. COMPACT($M + M$) =COMPACT(M)$\oplus$COMPACT(M).

where $\infty : \mathcal{C} \rightarrow \mathbb{R}^+ \cup \{\infty\}$ is defined by $\infty(c) = \infty$, for all $c \in \mathcal{C}$.

We now define the concatenation operator $\bowtie$:

Definition 16. The operator $\bowtie$ is defined by:

1. $\text{action}(g, c, \lambda) \bowtie [\![p]\!] = \text{action}(g, c, \lambda);[\![p]\!]$.

2. $\text{action}(g, c, \lambda) \bowtie \hat{M} = \text{action}(g, c, \lambda);\hat{M}$.

3. $\text{action}(g_1, c_1, \lambda_1) \bowtie \text{action}(g_2, c_2, \lambda_2) = \text{action}(g_1 \wedge g_2, c_1 \wedge c_2, \min(\lambda_1, \lambda_2))$.

where $\min(\lambda_1, \lambda_2) : \mathcal{C} \rightarrow \mathbb{R}^+ \cup \{\infty\}$ is defined by $\min(\lambda_1, \lambda_2)(c) = \min\{\lambda_1(c), \lambda_2(c)\}$.

Going back to the agent RW$_X$ previously defined; if we apply the function COMPACT to it, we obtain the following agent:

COMPACT(RW_X) :-
    action($X > 0, true, 1$) $\bowtie$ action($true, X = X - 1, \infty$) $\bowtie$ $[\![\text{RW}_X]\!]$
  $\oplus$ action($true, X = X + 2, 1$) $\bowtie$ $[\![\text{RW}_X]\!]$
  $\oplus$ action($true, true, f(X)$) $\bowtie$
    (action($X > 1, true, 1$) $\bowtie$ action($true, X = X - 2, \infty$) $\bowtie$ $[\![\text{RW}_X]\!]$
    $\oplus$ action($true, X = X + 1, 1$) $\bowtie$ $[\![\text{RW}_X]\!]$)

After the removal of $\bowtie$ operator according to the rules in Definition (16), the agent COMPACT(RW_X) becomes a COMPACT(sCCP) agent:

COMPACT(RW_X) = $[\![\text{RW}_X]\!]$:-
    action($true, X = X - 1, 1$) ; $[\![\text{RW}_X]\!]$
  $\oplus$ action($true, X = X + 2, 1$) ; $[\![\text{RW}_X]\!]$
  $\oplus$ action($true, true, f(X)$) ;
    (action($true, X = X - 2, 1$) ; $[\![\text{RW}_X]\!]$
    $\oplus$ action($true, X = X + 1, 1$) ; $[\![\text{RW}_X]\!]$)

The above example shows clearly that the function COMPACT simply collapses all the actions performed on the store after one execution of the stochastic transition, as defined in [25, 26]. It is a simple exercise to define a stochastic transition relation for COMPACT($sCCP$) similar to the one for RESTRICTED($sCCP$) and to successively prove the strong equivalence [128] between agents A and COMPACT(A).[4] Hence, from a semantic point of view, the application of function COMPACT is safe, as stated in the following

Lemma 9. *For each sequential agent A of* RESTRICTED($sCCP$)*, A and $\hat{A} =$ COMPACT(A) are strongly equivalent.*

Let $\hat{A} =$ COMPACT(A) be an agent of COMPACT($sCCP$). We want to associate a graph to such agent, containing all possible actions that $\hat{A}$ may execute. Nodes in such graph will correspond to different internal states of $\hat{A}$, i.e. to different stochastic branching points. Edges, on the other hand, will be associated to actions: each edge will correspond to one action(g, c, λ) instruction and will be labeled consequently by the triple (g, c, λ).

To define such graph, we proceed in two simple steps:

1. First we define an equivalence relation $\equiv_c$ over the set of COMPACT($sCCP$) agents, granting associativity and commutativity to $\oplus$ and reducing procedure calls to automatic "macro-like" substitutions (a reasonable move as we do not pass any parameter). We will then work on the set $\mathcal{A}$ of COMPACT($sCCP$) agents modulo $\equiv_c$, called the set of *states*; notably, all agents in $\mathcal{A}$ are stochastic summations.

2. Then, we define a structural operational semantics [192] on COMPACT($sCCP$), whose labeled transition system (LTS) will be exactly the target graph.

Definition 17. The equivalence relation $\equiv_c$ between COMPACT($sCCP$) agents is defined as the minimal relation closed with respect to the following three rules:

1. $\hat{M}_1 \oplus \hat{M}_2 \equiv_c \hat{M}_2 \oplus \hat{M}_1$;

2. $\hat{M}_1 \oplus (\hat{M}_2 \oplus \hat{M}_3) \equiv_c (\hat{M}_1 \oplus \hat{M}_2) \oplus \hat{M}_3$;

---

[4]Strong equivalence [128] is a form of bisimulation preserving probabilities: two agents are strongly equivalent if their exit rates are the same and transitions of one agent can be matched by transitions of the other having the same probability.

3. $[\![p]\!] \equiv_c \hat{A}$ if $[\![p]\!] : -\hat{A}$ belongs to the declarations $\hat{D}$.

The space of COMPACT($sCCP$) agents modulo $\equiv_c$ is denoted by $\mathcal{A}$, and is referred to as the *space of states*.

Definition 18. The transition relation $\leadsto \subseteq \mathcal{A} \times (\mathcal{C} \times \mathcal{C} \times \mathbb{R}^{\mathcal{C}}) \times \mathcal{A}$ is defined in the SOS style as the minimal relation closed with respect to the following rule:

$$\text{action}(g, c, \lambda); \hat{G} \oplus \hat{M} \stackrel{(g,c,\lambda)}{\leadsto} \hat{G}.$$

The transition relation $\leadsto$ encodes the possible actions that a COMPACT($sCCP$) agent can undertake. Notice that procedure calls are automatically solved as we are working modulo $\equiv_c$. The relation $\leadsto$ induces a labeled graph, its labeled transition system (LTS), whose nodes are agents in $\mathcal{A}$ and whose edges are labeled by triples (g, c, λ), where $g \in \mathcal{C}$ is a guard, $c \in \mathcal{C}$ is the update of the store, and λ the functional rate of the edge.

Definition 19. Let $\hat{A}$ be an agent of COMPACT($sCCP$); the portion of the labeled transition system reachable from the state $\hat{A}$ is denoted by $LTS(\hat{A})$.

Theorem 16. *For any agent $\hat{A}$ of COMPACT($sCCP$) (modulo $\equiv_c$), $LTS(\hat{A})$ is finite.*

Proof. The agents reachable from $\hat{A}$ are subagents of $\hat{A}$ or subagents of $\hat{A}'$, where $[\![p]\!] : -\hat{A}'$ is a procedure called by an agent reachable from $\hat{A}$. The number of subagents of $\hat{A}$ (modulo $\equiv_c$) corresponds to the number of summations present in $\hat{A}$, and it is finite for any definable agent. The proposition follows because there is only a finite number of agents defined in the declarations $\hat{D}$. ∎

Finiteness of the LTS implies also the computability of relation $\leadsto$, as its pairs can be determined simply by traversing the syntactic tree of each agent $\hat{A}$. We are finally ready to define the reduced transition system for an agent A of RESTRICTED(sCCP).

Definition 20. Let A be an agent of RESTRICTED(sCCP). Its *reduced transition system* $RTS(A)$ is a finite labeled multigraph $(S(A), T(A), \ell_A)$ defined by

$$RTS(A) = LTS(\text{COMPACT}(A)).$$

Given $RTS(A) = (S(A), T(A), \ell_A)$, $S(A) \subseteq \mathcal{A}$ is the set of *RTS-states* reachable from agent COMPACT(A), finite for Theorem 16, $T(A)$ is the set of *RTS-edges* or *RTS-transitions* and $\ell_A : T(A) \to \mathcal{C} \times \mathcal{C} \times \mathbb{R}^{\mathcal{C}}$ is the label function assigning to each RTS-edge the triple (g, c, λ), $g, c \in \mathcal{C}, \lambda : \mathcal{C} \to \mathbb{R}^+$.

Going back to our running example, $RTS(\text{RW}_X) = LTS(\text{COMPACT}(\text{RW}_X))$ is shown in the figure below. Note that it has one RTS-edge for every action that can be performed by COMPACT(RW_X), and just two RTS-states, corresponding to the two summations present in COMPACT(RW_X).

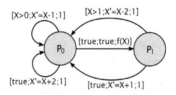

Step 2: the interaction matrix.

Our next step consists in encoding all the information about the dynamics in a single *interaction matrix* and in a *rate vector*. Consider the initial sCCP-network $N = A_1 \parallel \ldots \parallel A_h$ of a RESTRICTED($sCCP$) program $(Prog, \mathbf{X}, init(\mathbf{X}))$, with sequential components $A_1, \ldots, A_h$. First of all, we construct the reduced transition system for all the components, i.e. $RTS(A_1) = (S(A_1), T(A_1), \ell_{A_1}), \ldots, RTS(A_h) = (S(A_h), T(A_h), \ell_{A_h})$. Then we construct the set of RTS-states and RTS-transitions of the network (we agree that states and transitions belonging to different components $A_1, \ldots, A_h$ are distinct[5]), putting:

$$S(N) = S(A_1) \cup \ldots \cup S(A_h) \tag{9.1}$$

and

$$T(N) = T(A_1) \cup \ldots \cup T(A_h).[6] \tag{9.2}$$

Suppose now that there are m RTS-states in $S(N)$ and k RTS-transitions in $T(N)$. We conveniently fix an ordering of these two sets, say $S(N) = \{\sigma_1, \ldots, \sigma_m\}$ and $T(N) = \{t_1, \ldots, t_k\}$.

The variables $\mathbf{Y}$ of the differential equations are of two different kinds, $\mathbf{Y} = \mathbf{X} \cup \mathbf{P}$. The first type corresponds to the global stream variables of the store, i.e $\mathbf{X} = \{X_1, \ldots, X_n\}$ (see Definition 13). In addition, we associate a variable of the second type $P_{\sigma_i} = P_i$ to each RTS-state of $S(N) = \{\sigma_1, \ldots, \sigma_m\}$, so $\mathbf{P} = \{P_1, \ldots, P_m\}$. For the manipulations to follow, we assume the existence of a lexicographic ordering among all variables, so that vectors and matrices depending on this ordering are defined uniquely and manipulated consistently. Moreover, variables will be also used to index of these objects.

Consider now an RTS-transition $t_j \in T(N)$, connecting RTS-states σ_{j_1} and σ_{j_2}, and suppose $\ell_N(t_j) = (g_j, c_j, \lambda_j)$. We introduce the following notation:

- $\text{rate}_j^N(\mathbf{X}) = \lambda_j(\mathbf{X})$ is the rate function of t_j;

- $\text{guard}_j^N(\mathbf{X})$ is the indicator function of g_j (by Definition 13, g_j is a conjunction of linear equalities and inequalities), i.e.

$$\text{guard}_j^N(\mathbf{X}) = \begin{cases} 1 & \text{if } g_j \text{ is true for } \mathbf{X}, \\ 0 & \text{otherwise.} \end{cases} \tag{9.3}$$

We are now able to define the *rate vector*:

Definition 21. The *rate vector* r^N for transitions $T(N) = \{t_1, \ldots, t_k\}$ is a k-dimensional vector of functions, whose components r_j^N are defined by

$$r_j^N(\mathbf{Y}) = \text{rate}_j^N(\mathbf{X}) \cdot \text{guard}_j^N(\mathbf{X}) \cdot P_{j_1}, \tag{9.4}$$

where P_{j_1} is the variable associated to the source state σ_{j_1} of transition t_j.

Consider again a transition $t_j \in T(N)$, going from σ_{j_1} to σ_{j_2} and with label $\ell_N(t_j) = (g_j, c_j, \lambda_j)$, and consider the updates c_j, a conjunction of constraints of the form $X_i = X_i \pm k$, according to Definition 13. We can assume that each variable X_i appears in at most one conjunct of c_j.[7] We are now ready to define the *interaction matrix*

---

[5]If the same component is present in multiple copies, we distinguish among them by suitable labels.
[6]The labeling function acting on $T(N)$ will be denoted consistently by ℓ_N.
[7]If, for instance, both $X_i = X_i + k_1$ and $X_i = X_i + k_2$ are in c_j. Then we can replace these two constraints with $X_i = X_i + (k_1 + k_2)$.

Definition 22. The *interaction matrix* $I_{\mathbf{Y}}^N$ for an sCPP-network N with respect to variables $\mathbf{Y}$ is an integer-valued matrix with $n + m$ rows (one for each variable of $\mathbf{Y}$) and k columns (one for each RTS-transition $T(N)$), defined by:

1. If $\sigma_{j_1} \neq \sigma_{j_2}$, then $I_{\mathbf{Y}}^N[P_{j_1}, t_j] = -1$ and $I_{\mathbf{Y}}^N[P_{j_2}, t_j] = 1$.

2. If $X_h = X_h \pm k$ is a conjunct of c_j, then $I_{\mathcal{Y}}^N[X_h, t_j] = \pm k$.

3. All entries not set by points 1,2 above are equal to zero.

For the agent RW_X, the interaction matrix $I_{\mathbf{Y}}^{\mathrm{RW}_X}$ for the variables $\mathbf{Y} = \{X, P_0, P_1\}$ is:

$$I_{\mathbf{Y}}^{\mathrm{RW}_X} = \begin{pmatrix} -1 & +2 & 0 & -2 & +1 \\ 0 & 0 & -1 & +1 & +1 \\ 0 & 0 & +1 & -1 & -1 \end{pmatrix} \begin{matrix} (X) \\ (P_0) \\ (P_1) \end{matrix} \tag{9.5}$$

Similarly, the rate vector r^{RW_X} is

$$r^{\mathrm{RW}_X} = \begin{pmatrix} P_0\langle X > 0\rangle, & P_0, & f(X)P_0, & P_1\langle X > 1\rangle, & P_1, \end{pmatrix}^T, \tag{9.6}$$

where $\langle \cdot \rangle$ denotes the logical value of a formula (i.e., 1 if the formula is true, 0 otherwise).

Step 3: writing ODE's.

Once we have the interaction matrix, writing the set of ODE's is very simple: we just have to multiply matrix $I_{\mathbf{Y}}^N$ by the (column) rate vector r^N, in order to obtain the vector $ode_{\mathbf{Y}}^N$:

$$ode_{\mathbf{Y}}^N = I_{\mathbf{Y}}^N \cdot r^N. \tag{9.7}$$

Each row of the $ode_{\mathbf{Y}}^N$ vector gives the differential equation for the corresponding variable. Specifically, the equation for variable Y_i is

$$\begin{aligned} \dot{Y}_i &= ode_{\mathbf{Y}}^N[Y_i] = \sum_{j=1}^{k} I_{\mathbf{Y}}^N[Y_i, j] \cdot r_j^N(\mathbf{Y}) \\ &= \sum_{j=1}^{k} \left(I_{\mathbf{Y}}^N[Y_i, j] \cdot \mathrm{guard}_j(\mathbf{X}) \cdot \mathrm{rate}_j(\mathbf{X}) \cdot P_{j_1} \right) \end{aligned}$$

For instance, the set of ODE's associated to the agent RW_X is

$$\begin{cases} \dot{X} = P_0(2 - \langle X > 0\rangle) + P_1(1 - 2\langle X > 1\rangle) \\ \dot{P}_0 = -\frac{1}{X^2+1}P_0 + P_1(1 + \langle X > 1\rangle) \\ \dot{P}_1 = \frac{1}{X^2+1}P_0 + P_1(1 + \langle X > 1\rangle) \end{cases}$$

In order to solve a set of ODE's, we need to fix the *initial conditions*. The variables $\mathbf{Y} = \mathbf{X} \cup \mathbf{P}$ of $ode_{\mathbf{Y}}^N$ are of two distinct types: $\mathbf{P}$, denoting states of the reduced transition systems of the components, and $\mathbf{X}$, representing stream variables of the store. The initial conditions for $\mathbf{P}$ are easily determined: we set to one all the variables corresponding to the initial states of RTS of each component, and to 0 all the others. Regarding $\mathbf{X}$, instead, initial conditions are given in the formula $init(\mathbf{X})$ of Definition 13, specifying the values assigned to stream variables before starting the execution of the sCPP program.

Elimination of redundant state variables.

Consider an sCCP component A whose reduced transition system $RTS(A)$ has just one RTS-state. Then, the $ode_{\mathbf{Y}}^N$ vector of an sCCP-network N having A as one of its components will contain a variable corresponding to this RTS-state, say P_i, with equation $\dot{P}_i = 0$ and initial value $P_i(0) = 1$. Clearly, such variable is redundant, and we can safely remove it by setting $P_i \equiv 1$ in all equations containing it and by eliminating its equation from the $ode_{\mathbf{Y}}^N$ vector. From now on, we assume that this simplification has always been carried out. As an example, consider the following agent

$$A :\text{-} \text{tell}_{f_1(X)}(X = X + 1).A$$
$$+ \quad \text{tell}_{f_2(X)}(X = X - 1).A$$

Its RTS contains just one state, corresponding to the only summation present in it, with associated variable P. As the other variable of the agent is X, the vector $ode_{\{X,P\}}^A$ contains two equations, namely

$$\begin{pmatrix} \dot{X} \\ \dot{P} \end{pmatrix} = \begin{pmatrix} f_1(X)P - f_2(X)P \\ 0 \end{pmatrix}.$$

The simplification introduced above just prescribes to remove the equation for P, setting its value to 1 in the other equations; therefore we obtain

$$ode_{\{X\}}^A = (f_1(X) - f_2(X)).$$

Notice that the set of variables $\mathbf{Y}$ is updated consistently, i.e. removing the canceled variables from it.

We summarize the whole method just presented defining the following operator

Definition 23. Let N be the sCCP-network of an RESTRICTED($sCCP$) program. With

$$ODE(N)$$

we denote the vector $ode_{\mathbf{Y}}^N$ associated to N by the translation procedure previously defined, after applying the removal of state variables coming from network components with just one RTS-state.

Compositionality of the transformation operator

In order to clearly state formal properties of the transformation, we need a version of the $ODE(N)$ indicating explicitly the variables $\mathbf{X}$ for which the differential equations are given. In the following, the variables for the equations $ODE(N)$ are indicated by $VAR(ODE(N))$.

Definition 24. Let N be the sCCP-network of an RESTRICTED($sCCP$) program and let $\mathbf{Y} = VAR(ODE(N))$ and $ODE(N) = ode_{\mathbf{Y}}^N$. The ordinary differential equations of N with respect to the set of variables $\mathbf{X}$, denoted by $ODE(N, \mathbf{X})$, is defined as

$$ODE(N, \mathbf{X})[X_i] = \begin{cases} ode_{\mathbf{Y}}^N[Y_j] & \text{if } X_i = Y_j \in \mathbf{Y}, \\ 0 & \text{otherwise.} \end{cases}$$

The operations performed on $ODE(N)$ by $ODE(N, \mathbf{X})$ simply consist in the elimination of the equations of $ODE(N)$ for the variables not in $\mathbf{X}$, and in the addition of equations $\dot{X} = 0$ for all variables X in $\mathbf{X}$ but not in $\mathbf{Y}$. We can also associate a new interaction matrix $I_{\mathbf{X}}^N$ to $ODE(N, \mathbf{X})$, whose rows are derived according to Definition 24. As the set of RTS-transitions $T(N)$ is unaltered by $ODE(N, \mathbf{X})$, the equation $ODE(N, \mathbf{X}) = I_{\mathbf{X}}^N \cdot r^N$ continues to hold.

We can now prove the following theorem, stating compositionality of ODE operator.

Theorem 17. *Let N_1, N_2 be two sCCP-networks, and let $N = N_1 \parallel N_2$ be their parallel composition. If $\mathbf{Y_1} = VAR(ODE(N_1))$, $\mathbf{Y_2} = VAR(ODE(N_2))$, and $\mathbf{Y} = \mathbf{Y_1} \cup \mathbf{Y_2}$, then[8]*

$$ODE(N_1 \parallel N_2, \mathbf{Y}) = ODE(N_1, \mathbf{Y}) + ODE(N_2, \mathbf{Y}).$$

Proof. The components in $N_1 \parallel N_2$ are the components of N_1 plus the components of N_2. Therefore, the set of RTS-transitions (i.e. edges in the RTS of the components) $T(N_1 \parallel N_2)$ of $N_1 \parallel N_2$ is equal to $T(N_1) \cup T(N_2)$. As each column of $I_{\mathbf{Y}}^{N_1 \parallel N_2}$ is either a transition of $T(N_1)$ or a transition of $T(N_2)$, it clearly holds $I_{\mathbf{Y}}^{N_1 \parallel N_2}[Y_i, t_j] = I_{\mathbf{Y}}^{N_h}[Y_i, t_j]$ if $t_j \in T(N_h)$, $h = 1, 2$. The following chain of equalities then follows easily from the definitions:

$$
\begin{aligned}
ODE(N_1 \parallel N_2, \mathbf{Y})[Y_i] &= \sum_{t_j \in T(N_1 \parallel N_2)} I_{\mathbf{Y}}^{N_1 \parallel N_2}[Y_i, t_j] r_j^{N_1 \parallel N_2} \\
&= \sum_{t_j \in T(N_1)} I_{\mathbf{Y}}^{N_1}[Y_i, t_j] r_j^{N_1} + \sum_{t_j \in T(N_2)} I_{\mathbf{Y}}^{N_2}[Y_i, t_j] r_j^{N_2} \\
&= ODE(N_1, \mathbf{Y}) + ODE(N_2, \mathbf{Y}).
\end{aligned}
$$

∎

9.2.1   Preservation of Rate Semantics

In Section 7.1 we discussed how biochemical reaction with general kinetic laws can be modeled in sCCP. Given a list of reactions, the standard praxis in computational chemistry is that of building a corresponding differential (set of ODE's) or stochastic (CTMC) model. The definition of such models is *canonical*, and it is fully specified by the reaction list, see [239] for further details.

sCCP models of biochemical reactions are generated by associating an agent, call it *biochemical agent*, to each reaction in the list. These agents are rather simple: they can execute in one single way, namely an infinite loop consisting of activation steps, where the agents compete stochastically for execution, with rate given by the kinetic law specified in the reaction arrow, and update steps, in which the store is modified according to the prescriptions of the reaction. This bijective mapping between reactions and sCCP biochemical agents soon implies that the stochastic model for sCCP is identical to the continuous-time Markov chain generated in classical stochastic simulations with Gillespie algorithm [102], for instance like those obtainable with a program like Dizzy [200, 57].[9] A different question is wether the ODE's that are associated to an sCCP model of biochemical reactions coincide with the canonical ones. In the rest of the section, we show that this is indeed the case.

Following the approach by Cardelli in [49], we can then say that the translation from sCCP to ODE's *preserves the rate semantics*. The sense of this sentence is better visualized in Figure 9.1, graphically depicting the correspondence between stochastic and differential models of biochemical reactions and of the derived sCCP agents. Preservation of rate semantics essentially means that the arrows in the diagram commute.

As a matter of fact, in [49] the author deals only with mass action kinetics, due to the intrinsic properties of π-calculus (all definable rates are mass-action like). In our setting, instead, functional rates and the constraint store allow us to deal with arbitrary chemical kinetics, including also Michaelis-Menten and Hill ones (cf. [58]). In the following, we formally prove the equivalence of ODE's obtained from sCCP agents with the corresponding classical ones.

---

[8]Here "+" denotes the usual sum of vectors.

[9]Stochastic simulations with Michaelis-Menten and Hill rate functions have been considered, for instance, in [201].

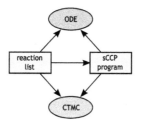

Figure 9.1: Diagram of relations intervening between stochastic and ODE-based semantics of chemical reactions and sCCP agents.

In general, the stochastic and the deterministic rate of a reaction are not the same, because ODE's variables measure concentration, while sCCP variables count the number of molecules. Therefore, in passing from one model to the other, we need to convert numbers to concentrations, dividing by the volume V times the Avogadro number N_A ($\gamma = V N_A$ will be referred as *system size*). Rates need also to be scaled consistently, see [239] for further details. In the rest of this section, however, we get rid of scaling problems simply by assuming $\gamma = 1$. In any case, system size can be reintroduced without difficulties, by change the scale of rates and variables after the derivation of ODE's.

Before going on any further, we recall from Chapter 7 how biochemical reactions are modeled in sCCP. Differently from Chapter 7, however, the approach presented here considers biochemical reactions parameterized w.r.t. their kinetics:

$$R_1 + \ldots + R_n \to_{f(\mathbf{R},\mathbf{X};\mathbf{k})} P_1 + \ldots + P_m. \tag{9.8}$$

The real-valued *kinetic function* of the reaction is $f(\mathbf{R}, \mathbf{X}; \mathbf{k})$, depending on the reactants $\mathbf{R}$, on other molecules $\mathbf{X}$ acting as modifiers, and on some parameters $\mathbf{k}$. This function can be one of the many used in biochemistry (cf. [58, 220] and Chapters 3 and 7) and it is required to satisfy the following *boundary condition*: it must be zero whenever one reactant is less than its amount consumed by the reaction. For instance, if a reactant R appears two times in the left hand size of (9.8), then f must be zero for $R = 0, 1$. [10].

In order to construct an sCCP model, to each reaction like (9.8), we associate the following sCCP agent:

$$f\text{-reaction}(\mathbf{R}, \mathbf{X}, \mathbf{P}, \mathbf{k}) :-$$
$$\text{tell}_{f(\mathbf{R},\mathbf{X};\mathbf{k})} \left(\bigwedge_{i=i}^{n} (R_i - 1) \wedge \bigwedge_{j=i}^{m} (P_j + 1) \right).$$
$$f\text{-reaction}(\mathbf{R}, \mathbf{X}, \mathbf{P}, \mathbf{k})$$

We now put forward some notation, in order to specify how to formally derive a set of ODE's given a set of reactions $\mathcal{R} = \{\rho_1, \ldots, \rho_k\}$, where each ρ_i denotes a single reaction. Each reaction ρ has some attributes: a multiset of reactants (species can have a specific multiplicity), denoted by $REACT(\rho)$, a multiset of products, $PROD(\rho)$, and a real-valued rate function, $RATE(\rho)$, depending on the variables associated to the molecules involved in the reaction, $VAR(\rho)$. This

---

[10]In case of mass action kinetics, this condition means that the rate for $R + R \to P$ must be $kR(R-1)$ and not kR^2. This is, however, consistent with the definition of the mass action principle in the stochastic setting.

last function, VAR, can be easily extended to sets of reactions by letting $VAR(\{\rho_1, \ldots, \rho_k\}) = VAR(\rho_1) \cup \ldots \cup VAR(\rho_k)$.

Now let $\mathcal{R}$ be a set of reactions and $\mathbf{X} = VAR(\mathcal{R})$. The (canonical) differential equations associated to $\mathcal{R}$ w.r.t. variables $\mathbf{X}$, denoted by $ODE(\mathcal{R}, \mathbf{X})$[11] are defined for each variable X_i as $\dot{X}_i = ODE(\mathcal{R}, \mathbf{X})[X_i]$, where:[12]

$$ODE(\mathcal{R}, \mathbf{X})[X_i] = \sum_{\substack{\rho \in \mathcal{R}: \\ X_i \in PROD(\rho)}} RATE(\rho) - \sum_{\substack{\rho \in \mathcal{R}: \\ X_i \in REACT(\rho)}} RATE(\rho). \qquad (9.9)$$

If X_i is not involved in any reaction of $\mathcal{R}$, then $ODE(\mathcal{R}, \mathbf{X})[X_i] = 0$. We observe that the correct stoichiometry is automatically dealt with by the fact that we are using multisets to list reactants and products. Restricting this construction to a fixed variable ordering allows to state the following straightforward compositionality lemma.

Lemma 10. *Let $\mathcal{R} = \mathcal{R}_1 \cup \mathcal{R}_2$ be a partition of $\mathcal{R}$ and let $\mathbf{X}_1 = VAR(\mathcal{R}_1)$, $\mathbf{X}_2 = VAR(\mathcal{R}_2)$, and $\mathbf{X} = \mathbf{X}_1 \cup \mathbf{X}_2$. Then*

$$ODE(\mathcal{R}_1 \cup \mathcal{R}_2, \mathbf{X}) = ODE(\mathcal{R}_1, \mathbf{X}) + ODE(\mathcal{R}_2, \mathbf{X}).$$

We turn now to formally define the encoding of reactions into sCCP agents. For each reaction ρ, its sCCP agent $SCCP(\rho)$ is constructed according to Section 7.1. Operator $SCCP$ is extended compositionally to sets of reactions $\mathcal{R} = \{\rho_1, \ldots, \rho_k\}$ by letting $SCCP(\mathcal{R}) = SCCP(\rho_1) \parallel \ldots \parallel SCCP(\rho_k)$.

We are finally ready to state the theorem of preservation of rate semantics:

Theorem 18 (Preservation of rate semantics). *Let $\mathcal{R}$ be a set of biochemical reactions, with $\mathbf{X} = VAR(\mathcal{R})$. Then*

$$ODE(\mathcal{R}, \mathbf{X}) = ODE(SCCP(\mathcal{R}), \mathbf{X}) \qquad (9.10)$$

Proof. We prove the theorem by induction on the size k of the set of reactions $\mathcal{R}$. For the base case $k = 1$, consider a reaction ρ

$$R_1 + \ldots + R_n \rightarrow_{f(\mathbf{R}, \mathbf{X}; \mathbf{k})} P_1 + \ldots + P_m$$

and its associated sCCP agent $SCCP(\rho)$

$$SCCP(\rho) :\text{-- } \text{tell}_{f(\mathbf{R}, \mathbf{X}; \mathbf{k})} \left(\bigwedge_{i=i}^{n} (R_i - 1) \wedge \bigwedge_{j=i}^{m} (P_j + 1) \right) .SCCP(\rho).$$

Clearly, the reduced transition system of such agent is

---

[11] We overload here the symbol introduced in Definitions 23 and 24; however, the two cases can be easily distinguished looking at their first argument.

[12] We conveniently identify each variable with the molecule it represents.

Let $\mathbf{Y} = VAR(\rho)$, then the interaction matrix $I_{\mathbf{Y}}^{SCCP(\rho)}$ has $|\mathbf{Y}|$ rows and 1 column, with entries corresponding to the stoichiometry of the reaction:

$$I_{\mathbf{Y}}^{SCCP(\rho)}[Y_i] = \sum_{Y_i \in PROD(\rho)} 1 - \sum_{Y_i \in REACT(\rho)} 1.$$

The rate vector $r^{SCCP(\rho)}$, instead, is the scalar $f(\mathbf{R}, \mathbf{X}; \mathbf{k})$. Hence, the equation for variable Y_i is

$$\dot{Y_i} = \sum_{Y_i \in PROD(\rho)} f(\mathbf{R}, \mathbf{X}; \mathbf{k}) - \sum_{Y_i \in REACT(\rho)} f(\mathbf{R}, \mathbf{X}; \mathbf{k}),$$

which is equal to equation (9.9).

The inductive case follows easily from compositionality properties of ODE operators. Suppose the theorem holds for lists up to $k-1$ reactions, and let $\mathcal{R} = \mathcal{R}_0 \cup \{\rho\}$ be a set of k chemical reactions (hence $|\mathcal{R}_0| = k - 1$). Then

$$
\begin{aligned}
ODE(SCCP(\mathcal{R}), \mathbf{X}) &= ODE(SCCP(\mathcal{R}_0 \cup \{\rho\}), \mathbf{X}) \\
&= ODE(SCCP(\mathcal{R}_0) \parallel SCCP(\rho), \mathbf{X}) \\
&= ODE(SCCP(\mathcal{R}_0), \mathbf{X}) + ODE(SCCP(\rho), \mathbf{X}) \\
&= ODE(\mathcal{R}_0, \mathbf{X}) + ODE(\rho, \mathbf{X}) \\
&= ODE(\mathcal{R}, \mathbf{X}),
\end{aligned}
$$

where the second equality follows from the definition of $SCCP$, the third follows from Theorem 17, the fourth is implied by the induction hypothesis on $SCCP(\mathcal{R}_0)$ and by the base case proof on $SCCP(\rho)$, while the last is a consequence of Lemma 10. ∎

9.2.2 Connections with the CTMC-based semantics

In this section we study more formally the relationships between the ODEs obtained from an sCCP program and its original semantics in terms of CTMC. We proceed in two steps. First, we define the so called Master Equation for the sCCP program. Secondly, we manipulate this equation, obtaining the ODEs generated by our method as a first-order approximation of the equations for the average of the CTMC associated to the sCCP program.

Master Equation for sCCP

The Master Equation (ME) is a (partial) differential equation for the time evolution of the probability mass distribution of a CTMC [137]. This equation is essentially equivalent to the Chapman-Kolmogorov forward equation (see [181] and Chapter 2), although it is more convenient to manipulate when the states of the system are vectors of (integer or real) variables whose evolution is determined by a fixed set of interactions. The ME is an exact equation, in the sense that it gives a precise picture of the evolution of the stochastic system. Unfortunately, it is almost impossible to solve, both analytically and numerically, hence approximate methods have been developed to deal with it.

The master equation for an sCCP network $N = A_1 \parallel \ldots \parallel A_n$ thus describes the time-variation of the probability mass function of the CTMC. First of all, we need to describe states of the CTMC by a set of variables. The choice is rather simple: a state of the network is described by the values of the stream variables $\mathbf{X}$ and by the variables associated to RTS-states $S(N)$, i.e. by a valuation of the variables $\mathbf{Y}$. Note that in this case we are not assuming these variables to be continuous, like in the ODE construction; in fact, they will generally be integer-valued.

Consider now the probability of being in state $\mathbf{Y}$ at time t, denoted by $\mathcal{P}(\mathbf{Y}, t)$. Each transition entering in state $\mathbf{Y}$ will increase this probability, while all transitions leaving $\mathbf{Y}$ will decrease it. Consider now a transition $t_j \in T(N)$. Its effect on $\mathbf{Y}$ is described by the j-th column vector $\nu_j = I_{\mathbf{Y}}^N[\cdot, j]$ of the interaction matrix $I_{\mathbf{Y}}^N$. In fact, the happening of t_j in $\mathbf{Y}$ will bring us in state $\mathbf{Y} + \nu_j$ (if $\mathbf{Y} + \nu_j$ does not belong to the domain of $\mathbf{Y}$, then transition t_j cannot happen — in this case, we assume its rate $r_j^N(\mathbf{Y})$ to be equal to zero). In addition, the probability that a transition t_j fires in the infinitesimal time dt, given that we are in state $\mathbf{Y}$, is $r_j^N(\mathbf{Y})dt$. The conditioning can be removed multiplying by $\mathcal{P}(\mathbf{Y}, t)$.

In summary, transition t_j increases $\mathcal{P}(\mathbf{Y}, t)$ in the infinitesimal time dt by

$$\mathcal{P}(\mathbf{Y} - \nu_j, t) r_j^N(\mathbf{Y} - \nu_j) dt,$$

and decrease it by

$$\mathcal{P}(\mathbf{Y}, t) r_j^N(\mathbf{Y}) dt.$$

The first term describes the probability of t_j leading into state $\mathbf{Y}$, while the second term describes the probability of t_j leading out of state $\mathbf{Y}$.[13]

Summing over all transitions and dividing for dt, we get the *master equation for the sCCP network N*:

$$\frac{\partial \mathcal{P}(\mathbf{Y}, t)}{\partial t} = \sum_j \left(r_j^N(\mathbf{Y} - \nu_j) \mathcal{P}(\mathbf{Y} - \nu_j, t) - r_j^N(\mathbf{Y}) \mathcal{P}(\mathbf{Y}, t) \right). \tag{9.11}$$

The initial conditions $\mathcal{P}(\mathbf{Y}, 0)$ for $\mathcal{P}$ are given by the initial configuration of the store (cf. Definition 13): at time 0 the system is in the state $\mathbf{Y_0}$ with probability 1, where the stream variables are determined by the constraints $X = x_0$, while the RTS-state variables are fixed by the initial state of each agent.

An important issue for the discussion to follow regards the continuity properties of rate functions r_j^N. Recalling their definition given in Definition 21, $r_j^N(\mathbf{Y}) = rate_j^N(\mathbf{X}) P_{j_1} guard_j^N(\mathbf{X})$, we note that r_j^N is the product of three functions: P_{j_1}, $rate_j^N$, and $guard_j^N$. P_{j_1} is a simple linear function (actually, a single variable), hence it is analytical. Generally, we can also expect $rate_j^N$ to be an analytical function (this is the case, for instance, for biochemical reactions). On the contrary, $guard_j^N$ is an indicator function, hence *discontinuous* whenever the guard $guard_j^N$ is non trivial (i.e., different from *true*). Guards are generally used to synchronize agents; in this sense, *discontinuity seems the price to pay for synchronization*. In some cases, however, the discontinuous nature of $guard_j^N$ is absorbed by the function $rate_j^N$. For instance, suppose that $guard_j^N(\mathbf{X}) = I(X_i > 0)$ and that $rate_j^N$ vanishes for $X_i = 0$; for all non-negative reals it then holds $guard_j^N rate_j^N = rate_j^N$, and so r_j^N is continuous whenever $rate_j^N$ is. This situation is not so uncommon, especially when guards are used to force upper and lower bounds on variables. An example are biochemical reactions, cf, below.

Consider now a set of biochemical reactions, whose sCCP encoding is given by a network of agents f-reaction, as in the previous section:

$$f\text{-reaction}(\mathbf{R}, \mathbf{P}, \mathbf{k}) :-$$
$$\quad \text{tell}_{f(\mathbf{R};\mathbf{k})}\left(\bigwedge_{i=i}^n (R_i - 1) \wedge \bigwedge_{j=i}^m (P_j + 1) \right).$$
$$\quad f\text{-reaction}(\mathbf{R}, \mathbf{P}, \mathbf{k}).$$

The function $f(\mathbf{R}; \mathbf{k})$, generally analytical (cf. Table 7.2 of Section 7.1), it is required to satisfy the following *boundary condition*: it must be zero whenever one reactant is less than its amount consumed by the reaction. For instance, if the reaction consumes two molecules of R, then f

---

[13]If $\mathbf{Y} - \nu_j$ does not belong to the domain of $\mathbf{Y}$, then we let $r_j^N(\mathbf{Y} - \nu_j) = 0$.

must be zero for $R = 0, 1$[14]. Note that the boundary conditions for the rate function f imply that no stream variable will ever become negative, as all reactions that may produce this effect have rate zero[15]. Consequently, we do not have to check domain constraints explicitly by guards, as their introduction would be redundant in the sense of the discussion at the end of previous section.

The RTS of this agent has just one state. In this case, the variable associated to it can be removed, as its equation would be $\dot{P} = 0$, $P(0) = 1$, and we can set $P \equiv 1$. Hence, the rate function φ associated to the agent f-reaction is $\varphi = f(\mathbf{R}; \mathbf{k})$. Now, the *chemical master equation* is defined as equation 9.11 with f in place of φ, hence the master equation of the sCCP model of a set of biochemical reactions coincides with the chemical master equation. The same result has been proved in [50] for chemical π-calculus, for mass action reactions with at most 2 reactants.

First-order approximation

From the Master Equation 9.11 we can deduce an exact differential equation for the average of each system variable Y_i of $\mathbf{Y}$. At time t, the average value $\mathbb{E}_t[Y_i]$ equals

$$\mathbb{E}_t[Y_i] = \sum_{\mathbf{Y}} Y_i \mathcal{P}(\mathbf{Y}, t),$$

where the sum ranges over all possible states of the system. As the only time-dependent quantity within the sum is $\mathcal{P}$, deriving both sides w.r.t. time t and substituting equation 9.11, we obtain

$$\frac{d\mathbb{E}_t[Y_i]}{dt} = \sum_j \left[\sum_{\mathbf{Y}} Y_i r_j^N(\mathbf{Y} - \nu_j) \mathcal{P}(\mathbf{Y} - \nu_j, t) - \sum_{\mathbf{Y}} Y_i r_j^N(\mathbf{Y}) \mathcal{P}(\mathbf{Y}, t) \right]. \quad (9.12)$$

In the first summation in the right-hand side, we can substitute $\mathbf{Y}$ with $\mathbf{W_j} + \nu_j$, breaking the sum in two pieces (one piece for W_{ij} and one piece for ν_{ij}). The first part cancels out with the second summation in 9.12, hence we are left with

$$\frac{d\mathbb{E}_t[Y_i]}{dt} = \sum_j \nu_{ij} \sum_{\mathbf{W_j}} r_j^N(\mathbf{W_j}) \mathcal{P}(\mathbf{W_j}, t).$$

Applying the definition of the average we obtain $\sum_{W_j} r_j^N(\mathbf{W_j}) \mathcal{P}(\mathbf{W_j}, t) = \mathbb{E}_t[r_j^N]$; recalling equation 9.7, we get

$$\frac{d\mathbb{E}_t[Y_i]}{dt} = \mathbb{E}_t[ode_{\mathbf{Y}}^N[Y_i](\mathbf{Y})] \quad (9.13)$$

Thus, the average value of $\mathbf{Y}$ in the stochastic system evolves as the *average* of the function $ode_{\mathbf{Y}}^N(\mathbf{Y})$, which is the vector of functions of the ODE system associated to the sCCP-network N by the fluid-flow approximation defined in Section 9.2.

Remark 1. If all rate functions r_j^N are *linear*, then so is $ode_{\mathbf{Y}}^N[Y_i](\mathbf{Y})$, and thus, thanks to the linearity of expectation $\mathbb{E}_t[\cdot]$, the equation 9.13 reduces to

$$\frac{d\mathbb{E}_t[Y_i]}{dt} = ode_{\mathbf{Y}}^N[Y_i](\mathbb{E}_t[\mathbf{Y}]),$$

---

[14]In case of mass action kinetics, this condition means that the rate for $R + R \to P$ must be $kR(R-1)$ and not kR^2. This is, however, consistent with the definition of the mass action principle in the stochastic setting.

[15]Boundary conditions for f may be relaxed by checking explicitly with ask instructions that variables stay within their domain. For instance, for the reaction $R + R \to P$, we can precede **tell** by **ask**$(R > 1)$. This allows us to use the more common kR^2 as rate function, at the price of introducing discontinuity in the rates.

which is equation 9.7. Therefore, if all rate functions r_j^N are linear, the method of Section 9.2 provides the exact equation for the average. This is essentially the same result proved in [78, 120].

Of course, if the linearity condition on r_j^N does not hold, then the previous remark is no more valid. In this case, the equation 9.7 is no more the correct equation for the average of the system. However, it is a first order approximation, hence reasonable whenever fluctuations are small. To see this, consider the time-dependent Taylor expansion of function $r_j^N(\mathbf{Y})$ around the average value $\mathbb{E}_t[\mathbf{Y}]$; for simplicity we truncate it at first order:

$$r_j^N(\mathbf{Y}) \approx r_j^N(\mathbb{E}_t[\mathbf{Y}]) + \sum_{k=1}^{|\mathbf{Y}|} \partial_k r_j^N(\mathbb{E}_t[\mathbf{Y}])(Y_k - \mathbb{E}_t[Y_k]),$$

where ∂_k denotes the partial derivative w.r.t. Y_k. Taking the average of both sides, the linear term cancels out, as $\mathbb{E}_t[(Y_k - \mathbb{E}_t[Y_k])] = 0$. Therefore, as $\mathbb{E}_t[ode_{\mathbf{Y}}^N(\mathbf{Y})] = \sum_j \nu_j \mathbb{E}_t[r_j^N(\mathbf{Y})]$, we obtain for $\mathbb{E}_t[ode_{\mathbf{Y}}^N(\mathbf{Y})]$ at first order:

$$\mathbb{E}_t[ode_{\mathbf{Y}}^N(\mathbf{Y})] \approx ode_{\mathbf{Y}}^N(\mathbb{E}_t[\mathbf{Y}]) \tag{9.14}$$

hence

$$\frac{d\mathbb{E}_t[Y_i]}{dt} \approx ode_{\mathbf{Y}}^N[Y_i](\mathbb{E}_t[\mathbf{Y}]).$$

The previous method works *only if* the functions r_j^N can be expanded in Taylor series (at least up to first order, hence r_j^N must have continuous first order derivatives). Preferably, r_j^N should be analytic. We stress that, in the presence of guards, this may not be true, even if all rates $rate_j^N$ are analytic.

9.2.3   Preservation of dynamic behavior

In Theorem 18 we proved that the *ODE* map, when applied to sCCP-models of biochemical networks, satisfies a condition of coherence: it preserves the kinetic principles used in the construction of the model (i.e., the rate semantics).

A different question is whether an sCCP-network N (evolving stochastically according to the prescriptions of its semantic) shows a dynamic behavior equivalent to the one exhibited by the equations $ODE(N)$. This problem is the sCCP-counterpart of the famous mathematical issue concerning the relation between stochastic and differential models [153, 154], studied deeply also in the context of biochemical reactions [102, 98]. It is well-known that stochastic and differential models of biochemical reactions are behaviorally equivalent only in some cases.

These results are significant also for sCCP. Theorem 18, in fact, states that, when biochemical reactions are concerned, the stochastic process underlying the sCCP-models and the associated ODE's are exactly the classical ones. Therefore, in the mapping from sCCP to ODE's we have the same phenomenology as in the classical case. However, the logical structure of sCCP-agents makes the problem of behavioral preservation subtler.

In fact, the discussion of Section 9.2.2 proved that the ODE map generates equations giving a first-order approximation of the average of the stochastic system, but only for synchronization free programs.

In the following, we discuss this problem with different examples, especially of situations in which an sCCP-network and the corresponding ODE's show a different behavior. In particular, we are interested in sketching a brief, and plausibly incomplete, classification of the causes of behavioral divergence.

An important issue is the concept of behavioral equivalence itself, which is difficult to formalize, as already discussed in the introduction. We will return on this problem in the next section.

Oregonator

The Oregonator is a chemical systems showing an oscillatory behavior, devised by Field and Noyes [182] as a simplified version of the Belousov-Zhabotinsky oscillator[16]. Essentially, Oregonator is composed of three chemical substances, call them A, B, C, subject to the following reactions:

$$
\begin{aligned}
B &\to_{k_1} A \\
A + B &\to_{k_2} \emptyset \\
A &\to_{k_3} 2A + C \\
2A &\to_{k_4} \emptyset \\
C &\to_{k_5} B
\end{aligned}
\tag{9.15}
$$

Actually, other chemical substances are involved, but they are kept constant in the experiment. The differential equations associated to (9.15) are known to possess a *stable limit cycle* for a wide range of parameter's values [119], containing an *unstable equilibrium*. The limit cycling behavior is clearly visible in Figure 9.2(a), where the numerical solution of Oregonator's ODE's is shown.

In Figure 9.2(b), instead, we plot a stochastic simulation of the sCCP model associated to (9.15) according to prescriptions of Section 7.1. In this case, the stochastic model shows the same pattern as the differential one. Theorem 18 guarantees that the graph in Figure 9.2 depicts the numerical solution of ODE's associated to the sCCP program by the transformation previously defined. In this case, the behavior is preserved.

We remark two things regarding Oregonator. First, the size of each molecular species is of the order of thousands, hence the relative variation induced by one reaction in the stochastic model is small. Under this condition, stochastic and deterministic models of biochemical reactions usually coincide [102]. Another property of the Oregonator that can be important for behavioral preservation is that the limit cycle is an *attractor* in the *phase space*: nearby trajectories asymptotically converge to it (see [227]). This means that a relatively small perturbation is not willing to change the overall dynamics: stochastic fluctuations have a negligible effect. Things are different if we start from the unstable equilibrium of the system. The numerical solution of ODE's shows a constant evolution (Figure 9.3(a)), while the stochastic simulation (Figure 9.3(b)) essentially evolves as the limit cycle of Figure 9.2. In fact, stochastic fluctuations, in this case, make the sCCP system move away from the instable equilibrium into the basin of attraction of the limit cycle. This shows another well known fact: *stochastic and differential models usually differ near instabilities* [102].

Lotka-Volterra system

The Lotka-Volterra system is a famous simple model of population dynamics, see for example [102] and references therein. There are two species: preys and predators. Preys eat some natural resource, supposed unbounded, and reproduce at a rate depending only on their number. Predators, instead, can reproduce only if they eat preys, otherwise they die. To keep the model simple, we admit predation as the only source of prey's death. The previous hypotheses can be summarized in the following set of reactions, where E refer to preys and C to predators:

$$
\begin{aligned}
E &\to_{k_b} 2E \\
C &\to_{k_d} \emptyset \\
E + C &\to_{k_p} 2C
\end{aligned}
\tag{9.16}
$$

---

[16]This chemical system is called "Oregonator" because its inventors where working at the University of Oregon.

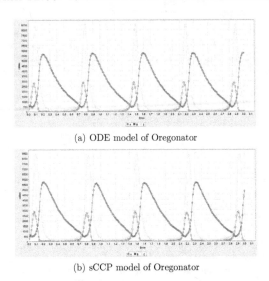

(a) ODE model of Oregonator

(b) sCCP model of Oregonator

Figure 9.2: **9.2(a)**: numerical simulation of the differential equation model of the Oregonator, with parameters determined according to the method presented in [102]. Specifically, let $A_s = 500$, $B_s = 1000$ and $C_s = 2000$ be an equilibrium of the system of equations, and let $R_1 = 2000$, $R_2 = 50000$. Then parameters are equal to $k_1 = R_1/B_s = 2$, $k_2 = R_2/(A_sB_s) = 0.1$, $k_3 = (R_1 + R_2)/A_s = 104$, $k_4 = ((2R_1)/(A_s^2))/2 = 4e^{-7}$, and $k5 = (R_1 + R_2)/C_s = 26$. The starting point is $A_0 = A_s/2$, $B_0 = B_s/2$, $C_0 = C_s/2$. The system soon approaches an attractive limit cycle. **9.2(b)**: stochastic simulation with Gillespie's method of the sCCP network associated to reactions (9.15). Parameters and initial conditions are those specified above. The effect of stochastic fluctuations is negligible, and the plot essentially coincide with its deterministic counterpart.

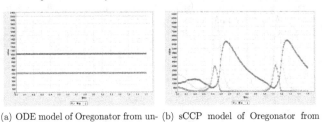

(a) ODE model of Oregonator from un- (b) sCCP model of Oregonator from
stable equilibrium   unstable equilibrium

Figure 9.3: **9.3(a)**: numerical simulations of ODE's derived from reactions (9.15), with parameters given in caption of Figure 9.2, starting from an unstable equilibrium of the system. **9.3(b)**: stochastic simulation of sCCP model associated to reactions (9.15), with the same parameters and initial conditions than the differential counterpart. As we can see, stochastic fluctuations drive the system away from the unstable equilibrium, so that its surrounding limit cycle is approached.

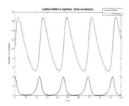 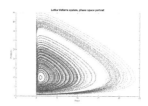

(a) ODE solution of Lotka-Volterra model (b) Phase space of Lotka-Volterra model

Figure 9.4: **9.4(a)**: numerical solution of ODE's associated to reactions (9.16), with parameters $k_b = 1$, $k_p = 0.1$, $k_d = 0.1$ and initial conditions $E_0 = 4$ and $C_0 = 10$. **9.4(b)**: phase portrait of the Lotka Volterra system, for the same value of parameters as above. As we can see, all the solutions show an oscillating behaviour. The system has an (instable) equilibrium for $E = 1$, $C = 10$, at the center of the circles.

If we consider the standard mass action ODE's (they coincide with the equations derived from the sCCP model due to Theorem 18), a typical solution shows oscillations in which high values of preys and predators alternate. An example of such a solution is given in Figure 9.4(a). Inspecting equations, it can be shown that the point $E_s = k_d/k_p$, $C_s = k_b/k_p$ is an equilibrium of a rather special kind: it is stable (trajectories starting nearby it stay close) but not asymptotically stable (trajectories starting nearby do not converge to it as time approaches infinity). This behavior is easily understood looking at the phase space (Figure 9.4(b)), in which we can see that trajectories form closed orbits around the equilibrium, whose amplitude increases with distance from equilibrium. More details can be found, for instance, in [227].

What kind of behavior can we expect from the stochastic evolution of the sCCP model for (9.16)? Stochastic fluctuations will make the system jump from one trajectory to nearby ones, without any force pulling it towards the equilibrium. Therefore, fluctuations can, in the long run, make the system wander in the phase plane, eventually reaching a borderline trajectory (corresponding to E or C axis in the phase plane). Whenever this happens, then both preys and predators go extinct (C-axis trajectory), or just predators do, while preys go to infinity (E-axis trajectory). This intuition is confirmed in Figure 9.5, where we compare the ODE solution starting from equilibrium (dotted lines), and a trace of the sCCP model, starting from the same initial configuration. As we can see, the stochastic system starts oscillating until both species go extinct.

This is another well known case in which stochastic and differential dynamics differ, again induced by properties of the phase space [102].

A negatively auto-regulated system

The effect of stochastic fluctuations is mostly remarkable in biological phenomena where gene expression is involved. This is because the transcription of a gene is usually a slower process than protein-protein interaction, and often the number of mRNA strands for a given gene present in the cell is very small, of the order of some units. As the production of one single mRNA is a rare event (compared to other cellular events), stochastic variability in its happening can induce behaviors difficult to capture if mRNA is approximated with its concentration. Stochasticity in gene expression is indeed a phenomenon that has received a lot of attention, see for instance [165, 21].

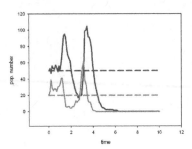

Figure 9.5: Effect of stochastic fluctuations for the Lotka-Volterra system. The dotted lines are an equilibrium solution for the ODE model (parameters are as in caption of Figure 9.4). A stochastic trace of the sCCP model is drawn with solid lines: both species fluctuate around the equilibrium values until they both get extinct.

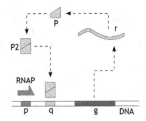

Figure 9.6: Diagram of a simple self-regulated gene network. Gene g produces mRNA r and, from it, protein P. P can dimerise and its dimer can bind to a promoter region of gene g, downstream of RNA polymerase binding site. The P_2-binding blocks polymerase activity, thus inhibiting gene expression.

We present here a simple, artificial example taken from [239] and depicted in Figure 9.6. The biological network shown represents a simple autoregulatory mechanism in gene expression of a procaryotic cell. Gene g produces, via mRNA r, a protein P that, as a dimer, can bind to a promoter region of gene g, preventing RNA-polymerase activity and thus inhibiting its own production.

Following the approach of [20], genes can be modeled as logical gates having a fixed output (the produced mRNA or protein), and several inputs, corresponding to different proteins of the system, exerting a positive or negative regulatory function. A gene gate with one inhibitory input is called in [20] *neg gate*, and can be modeled in RESTRICTED(sCCP) simply as:

$$
\begin{aligned}
\text{neg\_gate}_{P,I} :- \\
\quad \text{tell}_{k_p}(P = P + 1).\text{neg\_gate}_{P,I} \\
+ \quad \text{ask}_{k_b}.I(I \geq 1).\text{ask}_{k_u}(true).\text{neg\_gate}_{P,I},
\end{aligned}
$$

where k_p is the basic production rate, k_b is the binding rate of the repressor to the promoter region of the gene and k_u is its unbinding rate.

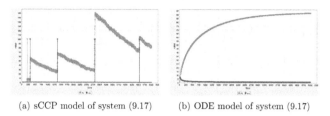

(a) sCCP model of system (9.17)   (b) ODE model of system (9.17)

Figure 9.7: **9.7(a)**: simulation of the sCCP model of (9.17). The red line corresponds to P_2, while the blue line shows the evolution of r, multiplied for a factor 100 (for visualization purposes). Note that the increases in P_2 expression immediately follow mRNA production events. Parameters of the models are the following: $k_p = 0.01$, $k_b = 1$, $k_u = 10$, $k_t = 10$, $k_{dim_1} = 1$, $k_{dim_2} = 1$, $k_{d_1} = 0.1$, and $k_{d_2} = 0.01$. All molecules are set initially equal to 0. **9.7(b)**: numerical simulation of ODE's associated to the sCCP model of (9.17), for the same parameters just given. The evolution of P_2 is tamer than in the stochastic counterpart, as it converges quickly to an asymptotic value.

In order to model the system of Figure 9.6 we can combine one neg gate with some reactions. This is an example of the modeling style mixing the reaction-centric and the molecular-centric point of view, see Chapter 7. The model is the following:

$$
\begin{aligned}
&\text{neg\_gate}_{r,P_2} \\
&r \to_{k_t} r + P \\
&P \to_{k_{dim_1}} P_2 \\
&P_2 \to_{k_{dim_2}} P \\
&r \to_{k_{d_1}} \emptyset \\
&P \to_{k_{d_2}} \emptyset
\end{aligned}
\qquad (9.17)
$$

In Figure 9.7 we compare a stochastic simulation of the sCCP model of reactions (9.17) with the numerical solution of the associated ODE's. As we can readily see, the two plots are completely different. In particular, in the stochastic simulation, P_2 is produced in short bursts; normally it is slowly degraded. The bursts correspond to mRNA production events, shown in Figure 9.7(a) as blue peaks. The ODE's system, however, presents a much simpler pattern of evolution, in which the quantity of P_2 converges to an asymptotic value. This divergence is caused by the fact that, approximating continuously the number of RNA molecules, we lose the discrete information that seems to characterize its dynamics, i.e. the fact that mRNA can be present in one unit of completely absent from the system.

Staten otherwise, continuously approximating molecular species present in low quantities may lead to errors inducing a completely divergent observable behavior.

Repressilator

The *Repressilator* [85] is an artificial biochemical clock composed of three genes expressing three different proteins, **tetR**, **λcI**, **LacI**, exerting a regulatory function on each other's gene expression. In particular, protein **tetR** represses the expression of protein **λcI**, protein **λcI** represses the gene producing protein **LacI**, and, finally, protein **LacI** is a repressor for protein **tetR**. The expected behavior is an oscillation of the concentrations of the three proteins. A simple stochastic model of Repressilator can be found in [20], where the authors describe it with three neg gates (see the previous paragraph) cyclically connected, in such a way that the product of one gate inhibits the successive gene gate in the cycle. In addition, they introduce degradation mechanisms for the three repressors. More formally, the model is the following

$$
\begin{aligned}
&\text{neg\_gate}_{A,C} \\
&\text{neg\_gate}_{B,A} \\
&\text{neg\_gate}_{C,B} \\
&A \to_{k_d} \emptyset \\
&B \to_{k_d} \emptyset \\
&C \to_{k_d} \emptyset
\end{aligned}
\tag{9.18}
$$

In Figure 9.8(a) we show a trace of the stochastic model generated by a simulator of sCCP based on Gillespie algorithm. The oscillatory behavior is manifest.

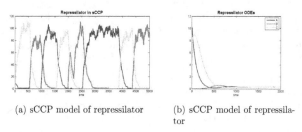

(a) sCCP model of repressilator

(b) sCCP model of repressilator

Figure 9.8: **9.8(a)**: stochastic time trace for the Repressilator system of described by reactions 9.18. Parameters are $k_p = 1$, $k_d = 0.01$, $k_b = 1$, $k_u = 0.01$. **9.8(b)**: solution of the differential equations of Table 9.3, automatically derived from sCCP program associated to reactions 9.18. Parameters are the same as in stochastic simulation.

If we apply the translation procedure discussed in Section 9.2 to this particular model, we obtain the ODE's shown in Table 9.3, while their numerical integration is shown in Figure 9.8(b). As we can readily see there is no oscillation at all, but rather the three proteins converge to an asymptotic value, after an initial adjustment.

Inspecting the ODE's, we note the presence of six variables (Y_A, Y_B, Y_C and Z_A, Z_B, Z_C) in addition to those representing the quantity of repressors in the system (A, B, C). Such variables correspond to states of genes gates, and they are used to model the change of configuration of the gates, from active to repressed and vice versa.

This scenario seems rather unjustified here: there is no argument to support the introduction of these variables, especially because we are continuously approximating boolean quantities.

An interesting point regarding Repressilator is the relation between the solution of the ODE's and the average trace of the stochastic system (i.e. $\mathbb{E}[\mathbf{X}(t)]$, returning the average value of system variables as a function of time). In fact, we may expect that the behavior preserved by the differential equations is the average dynamics of the stochastic system, rather than that shown by one of its traces. Interestingly, also the average value of the Repressilator model does not oscillate, as can be seen from Figure 9.9. This can be explained by noticing that the oscillations' period in the stochastic model is not constant, but it varies considerably. Hence, for every instant (when the Markov chain is at the stationary regime), we will observe one of the proteins at its peak value approximatively only in one third of the traces. Hence its average value will tend to stabilize at one third of the peak value, as confirmed by Figure 9.9. In fact, when we average Repressilator, we measure the fraction of traces in which a certain gene is

$$\dot{A} = k_p Y_A - k_d A \qquad \dot{Y}_1 = k_u Z_A - k_b Y_A C \qquad \dot{Z}_1 = k_b Y_A C - k_u Z_A$$
$$\dot{B} = k_p Y_B - k_d B \qquad \dot{Y}_2 = k_u Z_B - k_b Y_B A \qquad \dot{Z}_2 = k_b Y_B A - k_u Z_B$$
$$\dot{C} = k_p Y_C - k_d C \qquad \dot{Y}_3 = k_u Z_C - k_b Y_C B \qquad \dot{Z}_3 = k_b Y_C B - k_u Z_C$$

Table 9.3: ODE's derived for the Repressilator, generated by the method of Section 9.2.1.

active and the fraction of traces in which it is inactive, for every time instant. In this way, however, we lose any information regarding the sequence of gene gate's state changing. The different behavior existing between a trace of a stochastic system and its average trace suggests that the switching dynamics of genes can be the driving force behind oscillations. This implies that another source of non-equivalence between sCCP models and the associated ODE's can appear due to the representation of RTS-states with continuous RTS-state variables.

Indeed, this example suggested us to preserve part of the discrete dynamics, mapping the sCCP Repressilator into an hybrid automaton. The work put forward in [34] shows that this move is enough to maintain oscillations. The translation to hybrid automata opens an entire range of possibilities to combine discreteness and continuity. These will be investigated in detail in the planned second part of this paper.

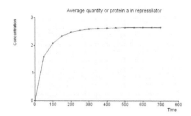

Figure 9.9: Average value of the sCCP model for Repressilator, computed using model checker PRISM [155]. See [26] for further details.

Sources of non-equivalence

In the previous examples we outlined different cases in which an sCCP model and its associated ODE's fail to be equivalent from a dynamical viewpoint. We remark that most of these examples are well known, as they have been studied in detail in theoretical and applicative contexts, like biochemical reactions [102, 239] and our main interest here is in their connection with the sCCP translation machinery. For sake of clarity, we summarize the different sources of non-equivalence.

1. In some cases, non-equivalence is a direct consequence of *properties of the phase space*. For instance, instable trajectories are destroyed by small fluctuations, like the equilibrium trajectory of the Oregonator. Also stable but not asymptotically stable trajectories can be troublesome, as stochastic fluctuations are not counterbalanced by any attracting force, and so they can bring the stochastic system far away from the initial trajectory. This is the case of the Lotka-Volterra system.

2. Another well-known problem is related to the *approximation by continuous quantities of integer variables having small (absolute) values*. In this case, in fact, the effect of a single stochastic fluctuation has a relative magnitude that is relevant, so the dynamics

can change quite dramatically. A typical example appearing in Biology is related to the transcription of genes, as shown in the simple example of a self-regulated gene.

3. A final source of non-equivalence is, instead, characteristic of the translation procedure defined for sCCP. In fact, in this case we *represent each RTS-state of a component of the system with a continuous variable*, which can take values in the real interval $[0, 1]$. RTS-states represent, in some sense, logical structures that control the activity of the system, while a change of state is an event triggered by some condition of the system. Moreover, in each sCCP trace, each component can be in only one state, hence RTS-state variables are boolean quantities. Continuous approximation, in this case, can have dramatic consequences, as the example of Repressilator seems to suggest.

9.2.4   Behavioral Equivalence

Comparing the dynamical evolution of a deterministic and a stochastic system is a delicate issue, because stochastic processes have a noisy evolution, hence we need to remove noise from their traces, before attempting any comparison with time traces evolving deterministically. In the previous discussion, in fact, we appealed to the concept of "behavioral equivalence" always in a vague sense, essentially leaving to the reader the task of visually comparing plots and recognizing similarities and differences. Clearly, a mathematical definition is needed in order to prove theorems and automatize comparisons.

We first consider the comparison of traces generated by ODE's with the average trace of the stochastic system, taken as the representative of its whole ensemble of traces. In practice, for each time instant t we need to compute the average value $\mathbb{E}(X(t))$ of each stream variable X w.r.t. the probability distribution on states of the system at time t. This probability can be obtained as the solution of the *Chapman-Kolmogorov forward equation* [181], a system of differential equations of the size of the state space. This equation, known in biochemical literature as the *chemical master equation* [101], can rarely be solved analytically, and it is also very difficult to integrate numerically [102]. A more efficient approach to compute an estimate of the average consists in generating several (thousands of) stochastic traces and in computing pointwise their sample mean. Alternatively, the average value of one or more variables can be computed for a small sample of time points $\{t_1, \ldots, t_k\}$ using numerical techniques, as those implemented in the model checker PRISM [155].

Whatever the method chosen, the computation (even approximate) of the average trace of a stochastic system is a difficult matter. Whenever such trace is known, we can compare it with the trace of the ODE's, generated using standard numerical techniques [193], using quantitative measures (essentially computing a distance between the two curves).

However, the average trace of a stochastic system is not necessarily a good representative of its evolution. A paradigmatic example is the Repressilator, whose average trace (sampled with PRISM, see caption of Figure 9.9) converges to an asymptotic value, while all its stochastic traces show persistent oscillations. Hence, *even when averaging a stochastic system, we may lose the characterizing qualitative features of its dynamics.*

The example of Repressilator suggests that the notion of behavioral equivalence is probably better captured in a qualitative setting. Qualitative comparison requires a formal definition of the *features* of dynamical evolution, like oscillations, convergence to a stable value, and so on. A possibility we suggest in this direction is to describe these features as logical formulae of a suitable logical language $\mathcal{L}$, for instance temporal logic, as done in Simpathica [9]. Let Φ denote the set of formulae describing all dynamical features of interest. Associating a Kripke structure K_1 to the trace of an ODE and another structure K_2 to a stochastic trace, then we may declare these traces equivalent whenever their Kripke structures satisfy the same subset of formulae of Φ (possibly restricting the attention to formulae of degree $\leq n$).

Below we give three examples of temporal logic formulas expressing infinite oscillations:

$$G(Z = z_m \rightarrow F(Z = z_M)) \ \wedge \ G(Z = z_M \rightarrow F(Z = z_m)) \ \wedge$$
$$G(z_m \leqslant Z \leqslant z_M) \ \wedge \ F(Z = z_m);$$

$$G(Z = z_m \rightarrow X(Z > z_m \ U \ Z = z_M)) \ \wedge \ G(Z = z_M \rightarrow X(Z < z_M \ U \ Z = z_m)) \ \wedge$$
$$G(z_m \leqslant Z \leqslant z_M) \ \wedge \ F(Z = z_m);$$

$$\left(\neg G \left(\frac{dZ}{dt} > 0 \right) \right) \ \wedge \ \left(\neg G \left(\frac{dZ}{dt} = 0 \right) \right) \ \wedge \ \left(\neg G \left(\frac{dZ}{dt} < 0 \right) \right).$$

In the above formulas G stands for *always (globally)*, F stands for *sometimes (in the future)*, U stands for *until*, z_m and z_M are minimum and maximum values, and the thirds formula uses propositional formulas taking values according to the sign of the first derivative.

This idea seems promising, as it gives a considerable freedom in the definition of formulae Φ, hence allowing to privilege some aspects of dynamical evolution more than others. However, the real problem is in the definition of a reasonable Kripke structure for a stochastic trace (and for sets of traces). In fact, Kripke structures for ODE's can be constructed starting from one or more traces, as done in Simpathica [9], in the following way: the bounded (product) domain of all variables is divided in small, compact regions; a state of the Kripke automaton consists of one of such regions; edges connect two states if a trajectory crosses the corresponding regions consecutively. This construction, however, is not reasonable for stochastic traces, as noise would force the addition of many edges that may introduce spurious behaviors. Of course, it is possible to model check directly on CTMC formulae written in CSL [11, 155]. However, the complexity of this latter approach makes the definition of non-deterministic Kripke structures interesting also for stochastic traces. We are currently investigating this direction, considering the introduction of a bounded form of memory to tame noise.

9.2.5 More on the restrictions of the language

The version of sCCP we used up to this point is restricted in several aspects, see Definition 13 in Section 9.1. Actually, these restrictions have been introduced in order to define in a reasonably simple way the mapping to ODE's. We discuss them in detail in the following.

First, in the language we constrain all the agents to be sequential, i.e. no occurrence of the parallel operator is allowed. Essentially, sequential agents are automata cooperating together, a property that allows to represent them graphically in a simple way. Indeed, this restriction is only apparent: we can always convert a non-sequential agent into a network of sequential ones using additional (stream) variables of the constraint store. The idea is simply that of identifying all the syntactically different terms that are stochastic choices, associating a new variable to each of them. These variables are used to count the number of copies of each term that are in parallel. Agents are consequently modified: all their transitions become self-loops in the RTSs and the variations induced in the number of terms by transitions are dealt with by updating the new state variables. Finally, rates are corrected by multiplying them by the multiplicity variable associated to the agent executing the corresponding transition. This is justified by the fact that in Markovian models, the global rate of a set of actions is computed by adding all basic rates together—ultimately, a consequence of the properties of the exponential distribution [181]. For instance, consider the agents x and y, defined by $x :- \text{tell}_1(true).(y \parallel y)$, $y :- \text{tell}_1(true).x + \text{tell}_1(true).\mathbf{0}$. They can be made sequential by introducing two variables, X

and Y, counting the number of copies of x and y respectively and by replacing x by x':- $\text{ask}_X(X > 0).\text{tell}_\infty(X' = X - 1 \wedge Y' = Y + 2).x'$ and y by y' :- $\text{ask}_Y(Y > 0).\text{tell}_\infty(X' = X + 1 \wedge Y' = Y - 1).y' + \text{ask}_Y(Y > 0).\text{tell}_\infty(Y' = Y - 1).y'$. Note that the same trick can be used to transform each sCCP-network into an equivalent network where each sequential agent has an RTS with one single state; indeed, this is done implicitly by the transformation itself. However, writing programs in this form is less natural.

Another syntactic restriction regards the definition of local variables. Actually, variables in ODE's have a global scope. Of course, any local variable can be made global by suitably renaming it. There is a problem, however, concerning the fact that at run-time we may generate an unbounded number of local variables. This implies that their use may lead to a set of ODE's with an infinite number of variables (although each equation will depend only on a finite number of them). The uprising of an infinite number of variables requires more complex mathematical techniques, and it prevents the use of standard numerical solvers.

Finally, the third class of restrictions regards the constraint store. The restriction to numeric variables is obviously necessary, as we are mapping to ODE's. The restriction on the admissible constraints for the updating of variables, on the other hand, is related to the fact that each update in a sCCP program needs to be considered as a flux acting on some variables. Indeed, even a simple update like $X' = 0$ is difficult to render within ODE's framework, as it is inherently discrete. A possible way out is to mix the continuous ingredient of ODE's with discreteness, mapping sCCP programs to hybrid automata [124]. Within this formalism, updates like $X' = 0$ are perfectly admissible: they are *resets* associated to discrete transitions. This point will be further discussed in the second part of this paper.

9.3 From ODE's to sCCP

In this section we first define a transformation $SCCP$, associating an sCCP network to a generic set of ordinary differential equation, and then we analyze its mathematical properties and the relation with the the map ODE defined in the previous section.

In this conversion from ODE's to sCCP, we approximate continuous quantities by discrete variables. Therefore, this mapping will depend on an additional parameter, the *step* δ, specifying the *granularity* of the approximation of continuous variables. The magnitude of δ has a strong impact on the preservation of dynamical behavior; this point will be the content of Section 9.3.2.

Consider a system of first order ODE's with n variables $\mathbf{x} = (x_1, \ldots, x_n)$:

$$\dot{\mathbf{x}} = \mathbf{f}(\mathbf{x}).$$

We write it explicitly as:

$$\begin{pmatrix} \dot{x}_1 \\ \vdots \\ \dot{x}_n \end{pmatrix} = \begin{pmatrix} \sum_{j=1}^{h_1} f_{1j}(x_1, \ldots, x_n) \\ \vdots \\ \sum_{j=1}^{h_n} f_{nj}(x_1, \ldots, x_n) \end{pmatrix} \tag{9.19}$$

In the sCCP program associated to the ODE's (9.19), we approximate the continuous variables $\mathbf{x}$ with discrete stream variables $\mathbf{X}$. Definition 13, however, requires variables $\mathbf{X}$ to have integer values. In order to set the size of the basic increment to an arbitrary step δ, we can change variables in (9.19), setting $\mathbf{x} = \delta\mathbf{X}$ and expressing $\mathbf{f}$ with respect to $\mathbf{X}$ (in this way, a unit increment of X_i corresponds to an increment of δ of x_i). The equation for X_i thus becomes

$$\dot{X}_i = \sum_{j=1}^{h_i} \frac{1}{\delta} f_{ij}(\delta\mathbf{X}).$$

By letting $F_{ij}(\mathbf{X}; \delta) = \frac{1}{\delta} f_{ij}(\delta \mathbf{X})$ and $F_i(\mathbf{X}; \delta) = \sum_{j=1}^{h_i} F_{ij}(\mathbf{X}; \delta)$, we can express the system of ODE's as $\dot{\mathbf{X}} = \mathbf{F}(\mathbf{X}; \delta)$. Expliciting the equations, we obtain:

$$
\begin{pmatrix} \dot{X}_1 \\ \vdots \\ \dot{X}_n \end{pmatrix} = \begin{pmatrix} \sum_{j=1}^{h_1} F_{1j}(X_1, \ldots, X_n; \delta) \\ \vdots \\ \sum_{j=1}^{h_n} F_{nj}(X_1, \ldots, X_n; \delta) \end{pmatrix} \tag{9.20}
$$

The translation to sCCP simply proceeds associating an agent to each differential equation of each variable of (9.20):

Definition 25. *Let $\dot{x} = \mathbf{f}(x)$ be a set of ODE's, written as in (9.20). Let $X_i \in \mathbf{X}$ and $\delta \in \mathbb{R}^+$. The agent $\mathrm{man}_{X_i, \delta}$ is defined as*[17]

$\mathrm{man}_{X_i, \delta}$:-
$\sum_{j=1}^{h_i} \mathrm{ask}_{|F_{ij}(X_1, \ldots, X_n; \delta)|}(F_{ij}(X_1, \ldots, X_n; \delta) > 0).\ \mathrm{tell}_\infty(X_i' = X_i + 1).\mathrm{man}_{X_i, \delta}$
$+\quad \mathrm{ask}_{|F_{ij}(X_1, \ldots, X_n; \delta)|}(F_{ij}(X_1, \ldots, X_n; \delta) < 0).\ \mathrm{tell}_\infty(X_i' = X_i - 1).\mathrm{man}_{X_i, \delta}$

The agent $\mathrm{man}_{X_i, \delta}$ is a big summation where each addend of the differential equation for X_i contributes with two branches: both have rate equal to the modulus of function F_{ij}, but one is active when $F_{ij} > 0$, and it increments the value of X_i, while the other is active when $F_{ij} < 0$, decrementing X_i.

In order to construct the sCCP network associated to a set of ODE $\dot{\mathbf{X}} = \mathbf{f}(\mathbf{X})$, we simply need to define an agent $\mathrm{man}_{X_i, \delta}$ for each variable of $\mathbf{X}$, putting these agents in parallel. We can render this procedure in the following *SCCP* operator:

Definition 26. *Let $\dot{x} = \mathbf{f}(x)$ be a set of ODE's, and let $\delta \in \mathbb{R}$, $\delta > 0$. The sCCP-network associated to $\mathbf{f}(x)$, with increment's step δ, indicated by $SCCP(\mathbf{f}(x), \delta)$, is*

$$
SCCP(\mathbf{f}(x_1, \ldots, x_n), \delta) = \mathrm{man}_{X_1, \delta} \parallel \ldots \parallel \mathrm{man}_{X_n, \delta}, \tag{9.21}
$$

with $\mathbf{x} = \delta \mathbf{X}$.

The initial conditions of the sCCP program, given by $init(\mathbf{X})$, are $\mathbf{X}(0) = \frac{1}{\delta} \mathbf{x_0}$, where $\mathbf{x_0}$ are the initial conditions of the ODE's.

Functional rates of sCCP are central in the definition of this translation: *each addend of a differential equation becomes a rate in a branch of an sCCP summation.* This is made possible only due to the freedom in the definition of rates, because differential equations considered here are general.

The possibility of having general rates in sCCP is intimately connected with the presence of the constraint store, which contains information external to the agents. This means that part of the description of interactions can be moved from the logical structure of agents to the functional form of rates. Common stochastic process algebras like stochastic π-calculus [194] or PEPA [128], on the other hand, have simple numerical rates and they rely just on the structure of agents (and on additivity of the exponential distribution [181]) to compute the global rate. This restricts severely the class of functional rates that they can model. Indeed, in a recent work [53] Hillston introduces general rates in PEPA essentially through the addition of information external to the model, an approach similar in spirit to sCCP.

---

[17]The name "man" stands for manager.

9.3.1   Invertibility properties

We turn now to study the relation between the two translations defined, i.e. ODE and $SCCP$. We start by showing that $(ODE \circ SCCP)$ returns the original differential equations.

Theorem 19. *Let* $\dot{\mathbf{x}} = \mathbf{f}(\mathbf{x})$ *be a set of differential equations, with* $\mathbf{x} = (x_1, \ldots, x_n)$ *and* $\mathbf{X} = \frac{1}{\delta}\mathbf{x}$. *Then*

$$ODE(SCCP(\mathbf{f}(\mathbf{x}), \delta), \mathbf{X}) = \mathbf{f}(\mathbf{x}).$$

Proof. From Definition 26 we know that $SCCP(\mathbf{f}(x_1, \ldots, x_n), \delta) = \mathrm{man}_{X_1,\delta} \parallel \ldots \parallel \mathrm{man}_{X_n,\delta}$, and by Theorem 17,

$$ODE(\mathrm{man}_{X_1,\delta} \parallel \ldots \parallel \mathrm{man}_{X_n,\delta}, \mathbf{X}) = ODE(\mathrm{man}_{X_1,\delta}, \mathbf{X}) + \ldots + ODE(\mathrm{man}_{X_n,\delta}, \mathbf{X}).$$

Moreover, an agent $\mathrm{man}_{X_i,\delta}$ modifies only the value of variable X_i (see Definition 25), hence the vector $ODE(\mathrm{man}_{X_i,\delta}, \mathbf{X})$ can have a non-zero entry only in correspondence of X_i (this follows from Definition 22 of the interaction matrix). This implies that the equation for X_i is exactly the one already present in $ODE(\mathrm{man}_{X_i,\delta}, \mathbf{X})$. To generate a set of ODE's from the agent $\mathrm{man}_{X_i,\delta}$, we have first to obtain its reduced transition system. It is easy to see that it has the form

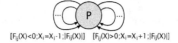

$$[F_{ij}(X)<0; X_i=X_i-1; |F_{ij}(X)|] \quad [F_{ij}(X)>0; X_i=X_i+1; |F_{ij}(X)|]$$

The interaction matrix derived from this RTS has only the row with non-zero entries, corresponding to variable X_i, each entry being equal to 1 or -1 (depending on the corresponding transition in the RTS). The corresponding ODE's is therefore

$$\dot{X}_i = \sum_{j=1}^{h_i} \left[|F_{ij}(\mathbf{X}; \delta)| \, \langle F_{ij}(\mathbf{X}; \delta) > 0 \rangle - |F_{ij}(\mathbf{X}; \delta)| \, \langle F_{ij}(\mathbf{X}; \delta) < 0 \rangle \right]. \qquad (9.22)$$

In the previous equation, $\langle \cdot \rangle$ denotes, as in Section 9.2, the logical value of a formula. Equation (9.22) can be simplified by noting that

$$|F_{ij}(\mathbf{X}; \delta)| \, \langle F_{ij}(\mathbf{X}; \delta) > 0 \rangle - |F_{ij}(\mathbf{X}; \delta)| \, \langle F_{ij}(\mathbf{X}; \delta) < 0 \rangle = F_{ij}(\mathbf{X}; \delta).$$

Hence, we obtain the equation $\dot{\mathbf{X}} = \mathbf{F}(\mathbf{X}; \delta)$, which is equal to $\dot{\mathbf{x}} = \mathbf{f}(\mathbf{x})$ when changing the variables back to $\mathbf{x}$.   ∎

Before turning the attention to dynamic properties, we take a closer look at what happens when we take the opposite point of view with respect to Theorem 19 and apply the operator $SCCP$ to a set of ODE's generated from an sCCP network. Consider the following two sCCP agents:

$$A :\text{-} (\mathrm{tell}_1(X = X + 1) + \mathrm{tell}_1(Y = Y + 1)).A$$

$$B :\text{-} (\mathrm{tell}_1(X = X + 1 \wedge Y = Y + 1)).B$$

When we apply the ODE operator to the networks $N_1 = A$ and $N_2 = B$, we obtain, in both cases, the following equations:

$$\begin{pmatrix} \dot{X} \\ \dot{Y} \end{pmatrix} = \begin{pmatrix} 1 \\ 1 \end{pmatrix}$$

Therefore, two different sCCP agents can be mapped into the same set of ODE's. Note that A and B are "semantically" different, as they induce two different CTMC. The chain associated to A has edges connecting a state (i, j) to $(i + 1, j)$ and $(i, j + 1)$ (hence the exit rate from (i, j) is 2), while the chain of B has transitions only from (i, j) to $(i + 1, j + 1)$ (with exit rate 1). An even worse situation happens for the following agent, implementing a one-dimensional random walk [181]:

$$C :\text{-} (\text{tell}_1(X = X + 1) + \text{tell}_1(X = X - 1)).C$$

The equation associated to C by ODE is $\dot{X} = 0$, as the production and degradation rate cancel out when summed together. This equation predicts a constant evolution for X, thus failing to capture its erratic behavior. Note, however, that the average value of X is constant also in the stochastic model for C.

From these examples we can conclude that in the translation from sCCP networks to differential equations we lose part of the logical structure of agents, due to the aggregation of rate functions given by the sum on reals. The structural information lost in passing from sCCP agents to ODE's makes impossible to recover the original sCCP network; staten otherwise, the map $ODE(\cdot)$ is not injective (but it is surjective, thanks to Theorem 19 applied with $\delta = 1$). Indeed, the lack of injectivity of $ODE(\cdot)$ means that an sCCP program is more informative than a set of ODE's: it defines not only the fluxes, but also their logical relation.

The previous discussion suggests a different way to consider the map $SCCP$: starting from a set of ODE's $\dot{\mathbf{x}} = \mathbf{f}(\mathbf{x})$, the map returns one sCCP-network belonging to the set $ODE^{-1}(\mathbf{f}(\mathbf{x}))$. A more general approach consists in requesting additional information to discriminate among the sCCP-networks of $ODE^{-1}(\mathbf{f}(\mathbf{x}))$: This information will essentially be related to the structure of the fluxes, hence to the logic of the system modeled.

9.3.2 Behavioral equivalence

We start this section by presenting an example showing how the translation from ODE's to sCCP works, with particular concern to the behavior exhibited by both systems and to the dependence on the step size δ, determining the size of the basic increment or decrement of variables. Intuitively, δ controls the "precision" of the sCCP agents w.r.t. the original ODE's. Varying the size of δ, we can *calibrate the effect of the stochastic fluctuations*, reducing or increasing it. This is evident in the following example, where we compare solutions of ODE's and the simulation of the corresponding sCCP processes.

Let's consider the following system of equations, representing another model of the Repressilator (see Section 9.2.3), a synthetic genetic network having an oscillatory behavior (see [85, 9]):

$$\begin{aligned} \dot{x}_1 &= \alpha_1 x_3^{-1} - \beta_1 x_1^{0.5}, & \alpha_1 = 0.2, & \quad \beta_1 = 0.01 \\ \dot{x}_2 &= \alpha_2 x_1^{-1} - \beta_2 x_2^{0.5}, & \alpha_2 = 0.2, & \quad \beta_2 = 0.01 \\ \dot{x}_3 &= \alpha_3 x_2^{-1} - \beta_3 x_3^{0.5}, & \alpha_3 = 0.2, & \quad \beta_3 = 0.01. \end{aligned} \tag{9.23}$$

The corresponding sCCP process, after changing variables according to $X_i = \frac{x_i}{\delta}$, is:

$$\text{man}_{X_1,\delta} \parallel \text{man}_{X_2,\delta} \parallel \text{man}_{X_3,\delta}, \tag{9.24}$$

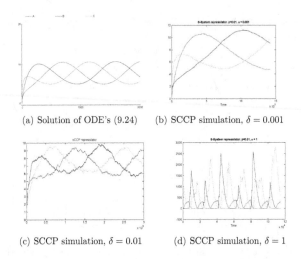

(a) Solution of ODE's (9.24)   (b) SCCP simulation, $\delta = 0.001$

(c) SCCP simulation, $\delta = 0.01$ (d) SCCP simulation, $\delta = 1$

Figure 9.10: Different simulations of sCCP agent obtained from S-Systems equations of repressilator (9.24), as basic step δ varies. Specifically, in Figure 9.10(a) we show the solution of ODE's (9.24), while in Figures 9.10(b), 9.10(c), 9.10(d) we present three simulations of the sCCP agent corresponding to ODE's (9.24), for $\delta = 0.001, 0.01, 1$ respectively. In the last diagram, the behavior of S-System's equations is destroyed.

where

$$\text{man}_{X_1,\delta} : -\big(\text{tell}_{[\frac{\alpha_1}{\delta}(\delta X_3)^{-1}]}(X_1' = X_1 + 1) + \text{tell}_{[\frac{\beta_1}{\delta}(\delta X_1)^{0.5}]}(X_1' = X_1 - 1)\big).\text{man}_{X_1,\delta}$$
$$\text{man}_{X_2,\delta} : -\big(\text{tell}_{[\frac{\alpha_2}{\delta}(\delta X_1)^{-1}]}(X_2' = X_2 + 1) + \text{tell}_{[\frac{\beta_2}{\delta}(\delta X_2)^{0.5}]}(X_2' = X_2 - 1)\big).\text{man}_{X_2,\delta}$$
$$\text{man}_{X_3,\delta} : -\big(\text{tell}_{[\frac{\alpha_3}{\delta}(\delta X_2)^{-1}]}(X_3' = X_3 + 1) + \text{tell}_{[\frac{\beta_3}{\delta}(\delta X_3)^{0.5}]}(X_3' = X_3 - 1)\big).\text{man}_{X_3,\delta}$$

In Figure 9.10, we study the dependence on δ of the sCCP network obtained from equations (9.24). From the plots, we note that the smaller δ, the closer to the solution of ODE's is the stochastic trace. However, increasing δ, the effect of stochastic perturbations gets stronger and stronger, making the system change dynamics radically.

Reducing the value of δ seems to be essentially the same as working with a sufficiently high number of molecules in standard biochemical networks, see [102, 99] and the discussion in Section 9.2.3. It is thus reasonable to expect that, by taking δ smaller and smaller, the deterministic and the stochastic dynamics will eventually coincide. In fact, reducing δ we are diminishing the magnitude of stochastic fluctuations, hence their perturbation effects.

This conjecture is indeed true: in the rest of the section we prove that, under mild conditions on the ODE's, the trajectories of the stochastic simulation converge to the solution of the ODE's. In fact, the set of stochastic traces whose distance from the solution of the ODE's is greater than a fixed arbitrary constant has zero probability in the limit $\delta \to 0$.

Kurtz theorem

In 1970 Thomas Kurtz proved a theorem giving conditions for a family of density dependent Continuous Time Markov Chains to converge to a solution of a system of ODE's [153, 154]. In fact, under mild assumptions on the smoothness of functions into play, the trajectories of the CTMC remain, in the limit, close to the solution of a particular set of ODE's with probability one. Our mapping $SCCP$ easily fits into Kurtz's framework, with the step δ playing the role of the density. We start by recalling the Kurtz's theorem.

Let V be a positive parameter, playing the role of the "size" of the system, and $\mathbf{X}_V(t)$ be a family of CTMC with state space $\mathbb{Z}^k$, depending on the parameter V. Suppose that there exist a continuous positive real function $\varphi : \mathbb{R}^k \times \mathbb{Z}^k \to \mathbb{R}$, such that the infinitesimal generator matrix [181] $Q = (q_{\mathbf{X},\mathbf{Y}})$ for $\mathbf{X}_V(t)$ is given by

$$q_{\mathbf{X},\mathbf{X}+\mathbf{h}} = V\varphi(\frac{1}{V}\mathbf{X}, \mathbf{h}), \quad \mathbf{h} \neq \mathbf{0}.$$

In addition, let $\Phi(\mathbf{x}) = \sum_{\mathbf{h}\in\mathbb{Z}^k} \mathbf{h}\varphi(\mathbf{x}, \mathbf{h})$.

Theorem 20 (Kurtz [153]). *Fix a bounded time interval $[0, T]$. Suppose there exists an open set $E \subseteq \mathbb{R}^k$ and a constant $M_E \in \mathbb{R}^+$ such that*

1. $|\Phi(\mathbf{x}) - \Phi(\mathbf{y})| < M_E|\mathbf{x} - \mathbf{y}|$, $\forall \mathbf{x}, \mathbf{y} \in E$ *(i.e. Φ satisfies the Lipschitz condition);*

2. $\sup_{\mathbf{x}\in E} \sum_{\mathbf{h}\in\mathbb{Z}^k} |\mathbf{h}|\varphi(\mathbf{x}, \mathbf{h}) < \infty;$

3. $\lim_{d\to\infty} \sup_{\mathbf{x}\in E} \sum_{|\mathbf{h}|>d} |\mathbf{h}|\varphi(\mathbf{x}, \mathbf{h}) = 0.$

Then, for every trajectory $\mathbf{x}(t)$ that is a solution of $\dot{\mathbf{x}} = \Phi(\mathbf{x})$ satisfying $\mathbf{x}(0) = \mathbf{x_0}$ and $\mathbf{x}(t) \in E$, $t \in [0, T]$, if

$$\lim_{V\to\infty} \frac{1}{V}\mathbf{X}_V(0) = \mathbf{x_0},$$

then for every $\varepsilon > 0$,

$$\lim_{V\to\infty} \mathbb{P}\left\{ \sup_{t\leq T} \left|\frac{1}{V}\mathbf{X}_V(t) - \mathbf{x}(t)\right| > \varepsilon \right\} = 0.$$

This theorem states that the trajectories of $\mathbf{X}_V(t)$ converge, in a bounded time interval, to the solution of $\dot{\mathbf{x}} = \Phi(\mathbf{x})$, when $V \to \infty$. The function Φ is essentially the sum of all fluxes of the system.

Convergence for $SCCP$

Our framework can be easily adapted to fit this theorem. Consider a system of ODE's $\dot{\mathbf{x}} = \mathbf{f}(\mathbf{x})$, and denote by $\mathbf{X}_\delta(t)$ the CTMC associated to the sCCP-network $SCCP(\mathbf{f}(\mathbf{x}), \delta)$.

Theorem 21. *Let $\dot{\mathbf{x}} = \mathbf{f}(\mathbf{x})$ be a system of ODE's, with $\mathbf{x} \in \mathbb{R}^k$, and $[0, T]$ a bounded time interval. If there exists an open set $E \subseteq \mathbb{R}^k$ such that $\mathbf{f}$ satisfies the Lipschitz condition in E and $\sup_{\mathbf{x}\in E} \|\mathbf{f}(\mathbf{x})\|_1 < \infty$, then for every $\varepsilon > 0$*

$$\lim_{\delta\to 0} \mathbb{P}\left\{ \sup_{t\leq T} |\delta\mathbf{X}_\delta(t) - \mathbf{x}(t)| > \varepsilon \right\} = 0,$$

where $\mathbf{x}(t)$ is the solution of $\dot{\mathbf{x}} = \mathbf{f}(\mathbf{x})$ with initial condition $\mathbf{x}(0) = \mathbf{x_0}$ and $\delta\mathbf{X}_\delta(0) = \mathbf{x_0}$.

Proof. In order to prove the theorem, we simply need to show that we satisfy all the hypothesis of the Kurtz's theorem. First of all, in this setting the density V is equal to $\frac{1}{\delta}$, so that $\frac{1}{\delta} \to \infty$ when $\delta \to 0$.
Consider now the ith function of $\mathbf{f}$, $f_i = \sum_j f_{ij}(\mathbf{x})$ and define $f_i^+(\mathbf{x}) = \sum_j f_{ij}(\mathbf{x}) \langle f_{ij}(\mathbf{x}) > 0\rangle$ and $f_i^-(\mathbf{x}) = \sum_j -f_{ij}(\mathbf{x})\langle f_{ij}(\mathbf{x}) > 0\rangle$, where $\langle\cdot\rangle$ denotes the logical value as before. Clearly,

$f_i(\mathbf{x}) = f_i^+(\mathbf{x}) - f_i^-(\mathbf{x})$ and $|f_i(\mathbf{x})| = f_i^+(\mathbf{x}) + f_i^-(\mathbf{x})$. Let also $\mathbf{e_1}, \ldots, \mathbf{e_k}$ be the canonical orthonormal basis of $\mathbb{R}^k$, so that $\mathbf{f}(\mathbf{x}) = \sum_i \mathbf{e_i} f_i(\mathbf{x})$.

Consider now the infinitesimal generator matrix $Q^\delta = (q_{\mathbf{X},\mathbf{Y}}^\delta)$ of the CTMC $\mathbf{X}_\delta(t)$. As the man$_{X_i, \delta}$ agents only increase or decrease a single variable by one unit (cf. Definition 25), it is straightforward to prove that

$$q_{\mathbf{X},\mathbf{X}+\mathbf{e_i}}^\delta = \frac{1}{\delta} f_i^+(\delta \mathbf{X}), \quad q_{\mathbf{X},\mathbf{X}-\mathbf{e_i}}^\delta = \frac{1}{\delta} f_i^-(\delta \mathbf{X}),$$

and all other entries of Q^δ are zero. Clearly, these are density dependent rates, with density $\frac{1}{\delta}$. Therefore, the function φ of the Kurtz theorem is simply defined as $\varphi(\mathbf{x}, \mathbf{e_i}) = f_i^+(\mathbf{x})$, $\varphi(\mathbf{x}, -\mathbf{e_i}) = f_i^-(\mathbf{x})$, and $\varphi(\mathbf{x}, \mathbf{h}) = 0$ for $\mathbf{h} \neq \pm\mathbf{e_i}$. Then, the function $\Phi(\mathbf{x})$ is $\Phi(\mathbf{x}) = \sum_{\mathbf{h}} \mathbf{h} \varphi(\mathbf{x}, \mathbf{h}) = \sum_i \mathbf{e_i}(f_i^+(\mathbf{x}) - f_i^-(\mathbf{x})) = \mathbf{f}(\mathbf{x})$.

It only remains to prove that conditions 1–3 of Theorem 20 are satisfied. Condition 1 is obvious because $\Phi = \mathbf{f}$ is Lipschitz by hypothesis, condition 3 hold because $|\mathbf{h}| > 1$ implies $\varphi(\mathbf{x}, \mathbf{h}) = 0$, while condition 2 follows because $\sum_{\mathbf{h}} |\mathbf{h}| \varphi(\mathbf{x}, \mathbf{h}) = \sum_i (f_i^+(\mathbf{x}) + f_i^-(\mathbf{x})) = \sum_i |f_i(\mathbf{x})| = \|\mathbf{f}(\mathbf{x})\|_1$ has a finite supremum in E by hypothesis. ∎

Comments and examples

Theorem 21 states that sCCP networks are able to simulate ODE's with an arbitrary precision. The cost of an exact stochastic simulation of the sCCP-network of Definition 26, however, is proportional to $\frac{1}{\delta}$, hence accurate stochastic simulations of ODEs are computationally impractical. On the other hand, there is no apparent reason to generate stochastic trajectories indistinguishable from the solution of the ODE's, as the latter can be generally obtained with much less computational effort.

In a work related to ours [95], Hillston et al. used the same Kurtz theorem to prove an analogous result for the equations that can be obtained from a PEPA program. Theorem 21 can be seen as a generalization of their result. Moreover, in [95] the authors suggest that a stochastic approximation of ODE's can be used together with analysis techniques typical of CTMC, like steady state analysis. This is a promising direction, but extreme care must be used.

Kurtz theorem, in fact, guarantees convergence only in a fixed and bounded time interval $[0, T]$, hence it does say nothing about asymptotic convergence of stochastic trajectories to ODE's.

Intuitively, the step δ may not be the only responsible for asymptotic convergence; an important role should also be played by initial conditions through topological properties of the phase space. If the ODE-trajectory we are considering is stable, i.e. resistant to small perturbations, then we can expect it to be reproduced in sCCP along the whole time axis, given a step δ small enough. On the other hand, if the trajectory is unstable, then even small perturbations can drive the dynamics far away from it; stochasticity, in this case, will unavoidably produce a trace dramatically different from ODE's one. Of course, by taking the interval $[0, T]$ of the theorem big enough (hence δ small enough), we can postpone arbitrarily far away in time the moment in which a stochastic and an unstable deterministic trajectories will diverge.

As an example of instability, let's consider a simple linear system of differential equations:

$$\begin{pmatrix} \dot{X} \\ \dot{Y} \end{pmatrix} = \begin{pmatrix} X + Y \\ 4X + Y \end{pmatrix} \tag{9.25}$$

The theory of dynamical systems [227] tells us that the point $(0, 0)$ is a *saddle node*, i.e. an unstable equilibrium whose phase space resembles the one depicted in Figure 9.11(a). The

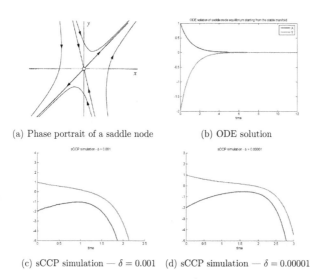

(a) Phase portrait of a saddle node   (b) ODE solution

(c) sCCP simulation — $\delta = 0.001$ (d) sCCP simulation — $\delta = 0.00001$

Figure 9.11: **9.11(a)**: phase space of the linear system (9.25). The origin is a saddle node; the stable manifold is displayed with arrows pointing towards the origin, while the unstable manifold has arrows diverging from it. **9.11(b)**: solution of the ODE's (9.25), starting from $(1, -2)$, a point belonging to the stable manifold. **9.11(c),9.11(d)**: simulation of the sCCP agent associated to the linear system (9.25), with initial conditions $(1, -2)$. The step δ is equal respectively to 0.001 and 0.00001. The time in which these trajectories diverge from the solution of the ODE's increases as δ becomes smaller.

two straight lines are the directions spanned by the eigenvectors of the matrix of coefficients $\begin{pmatrix} 1 & 1 \\ 4 & 1 \end{pmatrix}$, and are called *stable* and *unstable manifolds*. Motion in the stable manifold converges to the equilibrium $(0, 0)$, while the unstable manifold and all other trajectories diverge to infinity. However, small perturbations applied to the stable manifold can bring the system on a divergent hyperbolic trajectory, so we expect that ODE's and the associated sCCP agent, when starting from the stable manifold (say from point $(1, -2)$), will eventually jump on a divergent trajectory. Moreover, we expect that the smaller δ the later this event will happen. This intuition is confirmed in Figures 9.11(b), 9.11(c), 9.11(d).

This example shows that convergence issues need to be investigate further. In particular, conditions taking into account the topology of the phase space of the ODE's are required in order to guarantee also asymptotic convergence. Another interesting direction to investigate is to exploit other results by Kurtz [154] in order to state error bounds in the approximation.

9.4 Final discussion

In this chapter we presented a method to associate ordinary differential equations to sCCP programs (written with a restricted syntax), and also a method that generates an sCCP-network from a set of ODE's. The translation from sCCP to ODE's is based on the construction of a graph, called RTS, whose edges represent all possible actions performable by sCCP-agents. Properties of RESTRICTED($sCCP$) guarantee that the graph is always finite. From an RTS, we can construct an interaction matrix containing the modifications that each action makes to each variable. Writing the corresponding ODE's is simply a matter of combining the interaction matrix with the rate of each action. The inverse translation, from ODE's to sCCP, exploits the functional form that rates have in sCCP. In this way, we can associate sCCP-agents to general ODE's. An important feature of this method is that it is parametric w.r.t. the basic increment of variables, meaning that we can reduce the effect of stochastic fluctuations in the sCCP-model. Actually, we proved in Theorem 21 that, in the limit of an infinitesimal increment, the trajectories of the ODE's and of the corresponding sCCP-system coincide.

In Section 9.2.1, we showed that the translation from sCCP to ODE's, when applied to models of biochemical reactions, preserves the rate semantics in the sense of [49]. This condition, however, is not sufficient to guarantee that the translation maintains also the dynamical behavior of the sCCP-model. In fact, in Section 9.2.3, we provided several examples where an sCCP-network and the associated ODE's manifest a different behavior. This divergence can be caused by many factors, all qualitatively different.

Preserving dynamical behavior, however, is not just a mathematical game, but is is a central property that a translation from sCCP to ODE should have in order to be used as an analysis technique for stochastic process algebras. In this light, also the mapping from ODE to sCCP can be seen as a tool to investigate behavioral preservation.

The final part of Section 9.3, where the map ODE is shown not to be injective, suggests that in the passage to ODE's we unavoidably lose something of the sCCP model. This also suggest that the preservation of behavior may be reasonable only from a qualitative point of view. Indeed, this weaker approach fits better with the management of stochastic noise, see the discussion at the end of Section 9.2.3. The loss of precision in passing to ODE's is, however, counterbalances by the computational gain: simulating stochastic processes is undoubtedly much more expensive than numerically solving ODE's [99].

There are several open problems related to the question of behavioral equivalence. We list hereafter some of the most important ones, according to us.

- We need to identify the class of sCCP models (and their regions of parameter space/initial conditions) for which the mapping ODE preserves dynamics. Intuitively, according to discussion of Section 9.2.3, this may happen if all variables have big absolute values and if the phase space of the ODE's has asymptotically stable trajectories with ample basins of attraction.

- The repressilator and the simple self-inhibited genetic network of Section 9.2.3 suggest that the discrete ingredient cannot be continuously approximated so easily. In particular, associating continuous variables to RTS-states seems rather arbitrary. A possible solution can be that of transforming an sCCP network into a hybrid system, in which continuous and discrete dynamics coexist. In this way, we may be able to preserve part of the discrete structure of an sCCP-network, possibly just that fundamental for maintaining the behavior. We are investigating this direction, mapping sCCP-programs to hybrid automata [124, 5]. The first attempts in this direction are showing encouraging results [34, 35].

- The notion of behavioral equivalence needs to be specified formally. At the end of Section 9.2.3, we suggested an approach based on a suitable temporal logic, in which equivalence would mean equi-satisfiability of the same set of formulae.

As a final remark, we would like to consider this work under the perspective of the study of systemic properties. In fact, when we model a biological system, we are concerned mainly with the understanding of its systemic properties, especially what they are and how they emerge from basic interactions. In this direction, a modeler needs a formal language to specify biological systems, possibly provided with different semantics, related to one another and stratified in several layers of increasing approximation and abstraction. For example, sCCP has a natural CTMC-based semantics, but an ODE-based one can be assigned to it via the ODE operator. A possible layer in the middle consists in a semantic based, for instance, on hybrid automata. Finally, we need also a language to specify system's properties, automatically verifying them on the different semantics, or better, on the simpler semantic where answers are correct (i.e., on the simpler semantic showing the same dynamical behavior of the most general one). All these features must clearly be part of the same operative framework (and of the same software tool), hence all the open questions presented above can be seen as steps in this direction.

Conclusions

In this work we studied a stochastic extension of Concurrent Constraint Programming and used it as a modeling language to describe biological systems.

sCCP integrates two ingredients: the syntactic simplicity and the compositionality of process algebras on one side, and the descriptive and computational power of constraints on the other side. Therefore, we can describe separately the single pieces of biological systems we wish to model and then compose these parts together to create the final model. Moreover, constraints allow to introduce and manage complex information, like spatial one, in a fully extensible way, meaning that we can add new types of information simply defining suitable constraints, without the need of modifying the language itself. Hence, we can use sCCP to model biological systems with different degrees of complexity, from biochemical reactions to the folding process of a protein. sCCP can be thus be used as a general purpose language, whose constraint system can be instantiated time by time to the specific domain of the system to be modeled.

In biological modeling, the flexibility of the language reflects also on the point of view that the modeler can take while describing a class of systems. For instance, we can choose to describe biochemical reactions both taking an *entity-centric* view, modeling each molecule of the solution as a separate process, or a *reaction-centric* view, modeling each reaction as a process and each substance as a stream variable. We mainly chose this second approach, as it results more natural in sCCP. However, for other systems, like genetic regulatory networks, we used an hybrid approach, mixing *entity-centric* view (for genes) and *reaction-centric* view (for other protein activities).

In addition to the expressivity introduced by constraints, we have in sCCP another degree of flexibility, given by the use of non-constant rates in the definition of the stochastic duration of actions. For instance, these rates permit to use chemical kinetic theories different from the standard mass action. In addition, they give more power while translating sCCP to differential equations and viceversa: they the are that main ingredient allowing to associate sCCP programs to *general* ODEs.

The version of sCCP we presented here can be extended in different directions. For instance, we can add synchronous communication primitives, i.e. message exchanging on channels. The resulting language would therefore have instructions enabling a direct interaction between agents (synchronous communication), while keeping untouched the mechanisms for asynchronous indirect interaction via the constraint store. This improvement may be reasonable in the context of a distributed semantics of the language, where several computational locations may be defined so that processes could interact in the usual way within these loci and with message-passing across locations. This semantics may be defined following the approach of [37], possibly adding also features giving mobility and dynamism to computational nodes.

The implementation of the language we are using at the moment, i.e. the interpreter designed in Prolog, is not capable to simulate systems of big size. Hence, we need to implement a more efficient simulator, maybe incorporating also approximate simulation algorithms, like τ-leaping method [99].

Unfortunately, the dimension of real life systems make stochastic simulations a big computational issue. A possible solution would be that of using a distributed version of the Gillespie's simulation algorithm [229, 216]. Another possibility is that of modeling a cell by taking into account its compartmentalized structure. For instance, in order to take into account the slow motion of molecules due to the density of the cytoplasm, one can subdivide the cell into different blocks, letting molecules flow to neighboring blocks. A distributed semantics of sCCP may provide an interesting setting for this kind of modeling approach.

Stochastic model checking is undoubtedly a powerful alternative method of analysis that can be used to verify the properties of models. However, there is a limit in the size of Continuous Time Markov Chains that can be checked in a reasonable amount of time. A possible line of attack to extend these bounds may be that of associating to a stochastic model a non-deterministic automaton that may be analyzed more efficiently. Intuitively, one should introduce some form of memory in these automata, in order to let some probabilistic information survive. For instance, the set of traces of such hypothetic automaton should cumulate an high probability w.r.t the original Markov Chain. Stating the sentence otherwise, the non-deterministic automaton should preserve the behavior of the original system.

This last statement leads us directly to another open problem, i.e. that of defining a reasonable definition of behavioral equivalence, which may be used to compare models written in different languages or formalisms. A possibility in this direction can be that of using a temporal logic, considering the formulae that a model satisfies (maybe restricting the attention to special kinds of formulae related to dynamical behavior). In this way, two models are staten equivalent whenever they satisfy the same set of formulae. This definition may be relaxed by fixing a limit in the level of nesting of operators in the formulae considered.

This logical-based comparison can become an important tool in the study of equivalence between SPA and ODE models, shedding some light in the problem of defining translation procedures guaranteeing some form of behavioral invariance.

An important tool in this direction may be the use of hybrid automaton, especially in the direction from stochastic processes to ODEs. In fact, as the example of the repressilator of Chapter 9 clearly shows, the inner states of agents of stochastic models are not described well by continuous variables, but they may be described better by a mix of discrete states and non-determinism. Preliminary work in this direction showed interesting potentialities [34, 35].

Finally, another important issue is that of exploiting the computational power of the constraint store to its full extent. Its reasoning machinery, for instance, could be used to model dynamical compartments, diffusion, and many other phenomenon that must be kept into account in certain models.

The list of open problems just presented does not pretend to be exhaustive: there is a lot of work to do to make sCCP a competitive language in the difficult market of systems biology. However, we believe that it has a lot of potentialities to become a powerful modeling tool.

Appendix A

Simulated Annealing in sCCP

Simulated annealing is a very popular stochastic optimization algorithm, which has been presented in Section 5.2.1 of Chapter 5.

In Table A.1 we present a sequential version of the simulated annealing written in sCCP, where we have one single agent, `anneal`, implementing a recursive loop. The agent `anneal` first checks if the stop condition is satisfied, stopping the computation in the affirmative case. The stop condition is encoded in the constraint $stop(T, C)$, which can be defined according to one of the several strategies present in literature, see [3]. If the computation goes on, the agent calls the process `simulate`, implementing the moving strategy. Its first operation is the execution of the cooling strategy, hidden in the constraint $cooling(T)$.

After the temperature has been modified, a new candidate in the neighborhood of the current point is selected. The agent `choose_next`, which implements this mechanism, uses the constraint $neighbor(x, y)$ to verify if the point y is in the neighborhood of x. Note that this agent is a summation ranging over the whole space of points, and the ask guard guarantees that only those in the neighborhood of the current position can be effectively chosen. The rate function of these ask instructions gives equal weight to all branches, in such a way that the sum of the rates of active guards equals one. Note that we are assuming that our search space is finite. In the case of real functions, we can accomplish this by working with floating point numbers, or in interval arithmetics.

Finally, the agent `simulate` performs the usual Monte Carlo test of simulated annealing: the new point is accepted if it improves the energy, or with probability $e^{-\frac{f(x')-f(x)}{T}}$ in the other case. Also this mechanism, like the selection of a new candidate point, uses the stochastic semantics of sCCP: the test comparing the energies is performed by the ask guards, while the probabilistic choice is implemented using appropriate rate functions.

Note that the summation of agent `simulate` can have either the first branch active or the last two ones, but in each case the global rate of the summation equals one (the functions λ_a and λ_r sum punctually to one).

The variable C is a stream variable, like the current solution X, and it counts the number of moves attempted by the agent.

In Table A.2, we show how to adapt the general simulated annealing algorithm to the specific case of the multi-agent protein structure predictor presented in Chapter 8. In this case, the search algorithm is embodied in the process `update_position`, which extracts the current position of the aminoacid that called it, chooses randomly a new position and accepts it according to the usual Monte Carlo criterion. The variable $Newpos$ stores the position of the aminoacid at the end of the move. The constraint $assign\_position(I, \vec{P}, X_1)$ assigns to X_1, the coordinates of the Ith aminoacid, extracting them from the vector $\vec{P}$ containing the coordinates of all the amino acids.

---

start(X, T) :-
    $\exists_C$ (tell$_\infty(C\# = 1)$.anneal(X, T, C))

anneal(X,T,C) :-
        ask$_{\lambda_1}(stop(T, C))$.**0**
  $+$ ask$_{\lambda_1}(\neg stop(T, C))$.simulate$(X, T, C)$

simulate(X, T, C) :-
    tell$_\infty$(cooling(T)).
    $\exists_{X'}$(choose_next(X, X')).
    tell$_\infty(C\# = C + 1)$.
        ask$_{\lambda_1}(f(X')\# \le f(X))$.tell$_\infty(X\# = X')$.anneal$(X, T, C)$
  $+$ ask$_{\lambda_a(X,X',T)}(f(X')\# > f(X))$.tell$_\infty(X\# = X')$.anneal$(X, T, C)$
  $+$ ask$_{\lambda_r(X,X',T)}(f(X')\# > f(X))$.anneal$(X, T, C)$)

choose_next(X, X') :-
    $\sum_{\alpha \in S}$ ask$_{\lambda_N}$(neighbor(X, α)).tell$_\infty(X'\# = \alpha)$

$$\lambda_1 = 1$$

$$\lambda_a(X, X', T) = e^{\frac{f(X') - f(X)}{T}}$$

$$\lambda_r(X, X', T) = 1 - e^{\frac{f(X') - f(X)}{T}}$$

$$\lambda_1 = \frac{1}{n}, \quad \text{where } n = |\{\alpha : \text{neighbor}(X, \alpha)\}|$$

Table A.1: Sequential simulated annealing

update_position$(I, N, \vec{P}, Newpos)$:-
$\quad \exists_{X_1, X_2}($ tell$_\infty($assign_position$(I, \vec{P}, X_1))$
$\quad$ choose_next(X_1, X_2).
$\quad\quad$ ask$_{\lambda_1}(\Phi(I, X_2, \vec{P})\# \leq \Phi(I, X_1, \vec{P}))$.
$\quad\quad\quad$ tell$_\infty(Newpos\# = X_2)$.anneal(X, T, C)
$+\quad\quad$ ask$_{\lambda_a(X_1, X_2, T, \vec{P}, I)}(\Phi(I, X_2, \vec{P})\# > \Phi(I, X_1, \vec{P}))$.
$\quad\quad\quad$ tell$_\infty(Newpos\# = X_2)$.anneal(X, T, C)
$+\quad\quad$ ask$_{\lambda_r(X_1, X_2, T, \vec{P}, I)}(\Phi(I, X_2, \vec{P})\# > \Phi(I, X_1, \vec{P}))$.
$\quad\quad\quad$ tell$_\infty(Newpos\# = X_1)$.anneal(X, T, C))

choose_next(X, X') :-
$\quad \sum_{\alpha \in S}$ ask$_{\lambda_N}($neighbor$(X, \alpha))$.tell$_\infty(X'\# = \alpha)$

$$\lambda_1 = 1$$

$$\lambda_{a(X_1, X_2, T, \vec{P}, I)} = e^{\frac{\Phi(I, X_2, \vec{P}) - \Phi(I, X_1, \vec{P})}{T}}$$

$$\lambda_{r(X_1, X_2, T, \vec{P}, I)} = 1 - e^{\frac{\Phi(I, X_2, \vec{P}) - \Phi(I, X_1, \vec{P})}{T}}$$

$$\lambda_1 = \frac{1}{n}, \quad \text{where } n = |\{\alpha : \text{neighbor}(X, \alpha)\}|$$

Table A.2: Definition of the agent update_position, used in Chapter 8

Acknowledgments

Acknowledge people for their support in this work is not an easy task. I started working on these subjects back at the end of 2005, so almost three years passed, and three years can be a large amount of time. I'm pretty sure that I'll leave someone out, and I beg her or his pardon for this lapse.

I will start by thanking people which has been close to me with their love and their patience, namely my grandparents Augusto and Bruna, my parents Liviana and Luciano, and my irreplaceable partner Carla. Then my friends, of course, which I do not list so to avoid unforgettable omissions.

Many people worked side by side with me on these subjects, or simply helped me with precious hints and suggestions. I'll start by thanking Alberto Policriti, Agostino Dovier, and Andrea Sgarro who have been precious mentors in my apprenticeship as a researcher. Then, in sparse order, I owe more than something to Francesco Fabris, Alessandro dal Palù, Federico Fogolari, and Jane Hillston. I can't forget neither Eugenio Omodeo, Carla Piazza, Herbert Wikickly, Alessandra Di Pierro, Maria Grazia Vigliotti, Jeremy Bradley, Stephen Gilmore, Federica Ciocchetta, Vashti Galpin, Alberto Casagrande, Gabriele Puppis, Nicola Vitacolonna, Simone Scalabrin, Davide Bresolin, Marco Zantoni.

Bibliography

[1] Prism home page. http://www.cs.bham.ac.uk/~dxp/prism/.

[2] Converging sciences. Trento, 2004. http://www.unitn.it/events/consci/.

[3] E. Aarts and J. Korst. *Simulated Annealing and Boltzmann machines*. John Wiley and sons, 1989.

[4] E. H. L. Aarts and J. K. Lenstra, editors. *Local Search in combinatorial Optimization*. John Wiley and Sons, 1997.

[5] R. Alur, C. Belta, F. Ivancic, V. Kumar, M. Mintz, G. Pappas, H. Rubin, and J. Schug. Hybrid modeling and simulation of biomolecular networks. In *Proceedings of Fourth International Workshop on Hybrid Systems: Computation and Control*, volume LNCS 2034, pages 19–32, 2001.

[6] P. Amara, J. Ma, and J. E. Straub. *Global Minimization of Nonconvex Energy Functions: Molecular Conformation and Protein Folding*, chapter Global minimization on rugged energy landscapes, pages 1–15. Amer. Math. Soc., 1996.

[7] C.B. Anfinsen. Principles that govern the folding of protein chain. *Science*, 181:223–230, 1973.

[8] N. Angelopoulos, A. Di Pierro, and H. Wiklicky. Implementing randomised algorithms in constraint logic programming. In *Proceedings of Joint International Conference and Symposium on Logic Programming (JICSLP'98)*, 1998.

[9] M. Antoniotti, A. Policriti, N. Ugel, and B. Mishra. Model building and model checking for biochemical processes. *Cell Biochemistry and Biophysics*, 38(3):271–286, 2003.

[10] K. Apt. *Principles of Constraint Programming*. Cambridge University Press, 2003.

[11] A. Aziz, V. Singhal, F. Balarin, R. Brayton, and A. Sangiovanni-Vincentelli. Verifying continuous time markov chains. In *Proceedings of CAV96*, 1996.

[12] T. Bäck, D.B. Fogel, and Z. Michalewicz, editors. *Handbook of Evolutionary Computation*. IOP Publishing Ltd and Oxford University Press, 1997.

[13] J.C.M. Baeten. A brief history of process algebra. Technical Report CSR 04-02, Technische Universiteit Eindhoven, 2004.

[14] C. Baier, J. P. Katoen, and H. Hermanns. Approximate symbolic model checking of continuous time markov chains. In *Proceedings of CONCUR99*, 1999.

[15] Gianfranco Balbo. Introduction to stochastic petri nets. pages 84–155, 2002.

[16] H. M. Berman et al. The protein data bank. *Nucleic Acids Research*, 28:235–242, 2000. http://www.rcsb.org/pdb/.

[17] M. Berrera, H. Molinari, and F. Fogolari. Amino acid empirical contact energy definitions for fold recognition. *BMC Bioinformatics*, 4(8), 2003.

[18] P. Billingsley. *Probability and Measure*. John Wiley and Sons, 1979.

[19] M.L. Blinov, J. Yang, J.R. Faeder, and W.S. Hlavacek. Graph theory for rule-based modeling of biochemical networks. *Transactions of Computational Systems Biology*, 2007.

[20] R. Blossey, L. Cardelli, and A. Phillips. A compositional approach to the stochastic dynamics of gene networks. *T. Comp. Sys. Biology*, pages 99–122, 2006.

[21] R. Blossey, L. Cardelli, and A. Phillips. Compositionality, stochasticity and cooperativity in dynamic models of gene regulation. *HFPS Journal, in print*, 2007.

[22] A. Bockmayr and A. Courtois. Using hybrid concurrent constraint programming to model dynamic biological systems. In Springer-Verlag, editor, *Proceedings of 18th International Conference on Logic Programming, ICLP'02*, volume 2401 of *LNCS*, pages 85–99, 2002.

[23] F.S. de Boer, R.M. van Eijk, W. van der Hoek, and J-J.Ch. Meyer. Failure semantics for the exchange of information in multi-agent systems. In *Proceedings of CONCUR 2000*, 1998.

[24] R. Bonneau and D. Baker. Ab initio protein structure prediction: progress and prospects. *Annu. Rev. Biophys. Biomol. Struct.*, 30:173–89, 2001.

[25] L. Bortolussi. Stochastic concurrent constraint programming. In *Proceedings of 4th International Workshop on Quantitative Aspects of Programming Languages (QAPL 2006)*, volume 164 of *ENTCS*, pages 65–80, 2006.

[26] L. Bortolussi. *Constraint-based approaches to stochastic dynamics of biological systems*. PhD thesis, PhD in Computer Science, University of Udine, 2007. Available at http://www.dmi.units.it/~bortolu/files/reps/Bortolussi-PhDThesis.pdf.

[27] L. Bortolussi. A master equation approach to differential approximations of stochastic concurrent constraint programming. In *Submitted to QAPL 2008. Available upon request*, 2008.

[28] L. Bortolussi, A. Dal Palù, A. Dovier, and F. Fogolari. Protein folding simulation in CCP. In *Proceedings of BioConcur2004*, 2004.

[29] L. Bortolussi, A. Dal Palù, A. Dovier, and F. Fogolari. Agent-based protein folding simulation. *Intelligenza Artificiale*, January 2005.

[30] L. Bortolussi, A. Dovier, and F. Fogolari. Multi-agent simulation of protein folding. In *Proceedings of MAS-BIOMED 2005*, 2005.

[31] L. Bortolussi, S. Fonda, and A. Policriti. Constraint-based simulation of biological systems described by molecular interaction maps. In *Proceedings of IEEE conference on Bioinformatics and Biomedicine, BIBM 2007*, 2007.

[32] L. Bortolussi and A. Policriti. Relating stochastic process algebras and differential equations for biological modeling. *Proceedings of PASTA 2006*, 2006.

[33] L. Bortolussi and A. Policriti. Stochastic concurrent constraint programming and differential equations. In *Proceedings of Fifth Workshop on Quantitative Aspects of Programming Languages, QAPL 2007*, volume ENTCS 16713, 2007.

[34] L. Bortolussi and A. Policriti. Hybrid approximation of stochastic concurrent constraint programming. In *To be presented at IFAC 2008. Available upon request*, 2008.

[35] L. Bortolussi and A. Policriti. The importance of being (a little bit) discrete. In *Proceedings of FBTC'08*, 2008.

[36] L. Bortolussi and A. Policriti. Modeling biological systems in concurrent constraint programming. *Constraints*, 13(1), 2008.

[37] L. Bortolussi and H. Wiklicky. A distributed and probabilistic concurrent constraint programming language. In M. Gabbrielli and G. Gupta, editors, *Lecture Notes in Computer Science 3368, Proceedings of 21st International Conference of Logic Programming, ICLP 2005*, pages 143–158. Springer-Verlag, October 2005.

[38] Luca Bortolussi, Agostino Dovier, and Federico Fogolari. Multi-agent protein structure prediction. *International Journal on Multi-Agents and Grid Systems*, 3(2), 2007.

[39] J. M. Bower and H. Bolouri eds. *Computational Modeling of Genetic and Biochemical Networks*. MIT Press, Cambridge, 2000.

[40] P. Bradley, L. Malstrom, B. Qian, J. Schonbrun, D. Chivian, D. E. Kim, J. Meiler, K. Misura, and D. Baker. Free modeling with Rosetta in CASP6. *Proteins*, 7S:128–134, 2005.

[41] T. Brodmeier and E. Pretsch. Application of genetic algorithms in molecular modeling. *J. Comp. Chem.*, 15:588–595, 1994.

[42] R. A. Broglia and G. Tiana. Reading the three-dimensional structure of lattice model-designed proteins from their amino acid sequence. *Proteins: Structure, Functions and Genetics*, 45:421–427, 2001.

[43] B. R. Brooks et al. Charmm: A program for macromolecular energy minimization and dynamics. *J. Comput. Chem.*, 4:187–217, 1983.

[44] M. Calder, S. Gilmore, and J. Hillston. Modelling the influence of rkip on the erk signalling pathway using the stochastic process algebra PEPA. *Transactions on Computational Systems Biology*, 4230:1–23, 2006.

[45] L. Calzone, F. Fages, and S. Soliman. Biocham: an environment for modeling biological systems and formalizing experimental knowledge. *Bioinformatics*, 22:1805–1807, 2006.

[46] F. Capra. *The Web of Life*. Doubleday, New York, 1996.

[47] L. Cardelli. Abstract machines of systems biology. *Transactions on Computational Systems Biology*, III, LNBI 3737:145–168, 2005.

[48] L. Cardelli. From processes to odes by chemistry. *downloadable from* `http://lucacardelli.name/`, 2006.

[49] L. Cardelli. On process rate semantics. *draft*, 2006.

[50] L. Cardelli. A process algebra master equation. In *Proceedings of QEST 2007*, 2007.

[51] L. Cardelli and A. Phillips. A correct abstract machine for the stochastic pi-calculus. In *Proceeding of Bioconcur 2004*, 2004.

[52] N. Carriero and D. Gelernter. Linda in context. *Communications of the ACM*, 32(4):444–458, 1989.

[53] F. Ciocchetta and J. Hillston. Bio-PEPA: an extension of the process algebra PEPA for biochemical networks. In *Proceeding of FBTC 2007*, 2007. Workshop of CONCUR 2007.

[54] F. Ciocchetta, C. Priami, and P. Quaglia. Modeling kohn interaction maps with beta-binders: an example. *Transactions on computational systems biology*, LNBI 3737:33–48, 2005.

[55] E. Clarke, A. Peled, and A. Grunberg. *Model Checking*. MIT press, 1999.

[56] M. Clerc and J. Kennedy. The particle swarm: Explosion, stability, and convergence in a multi-dimensional complex space. *IEEE Journal of Evolutionary Computation*, 2001.

[57] Seattle CompBio Group, Institute for Systems Biology. Dizzy home page.

[58] A. Cornish-Bowden. *Fundamentals of Chemical Kinetics*. Portland Press, 3rd edition, 2004.

[59] D. G. Covell and R. L. Jernigan. Conformations of folded proteins in restricted spaces. *Biochemistry*, 29:3287–3294, 1990.

[60] T.G. Crainic and M. Toulouse. Parallel metaheuristics. *Technical report, Centre de recherche sur les transports, Université de Montreal*, 1997.

[61] P. Crescenzi, D. Goldman, C. Papadimitrou, A. Piccolboni, and M. Yannakakis. On the complexity of protein folding. In *Proc. of STOC*, pages 597–603, 1998.

[62] P.L. Curien, V. Danos, J. Krivine, and M. Zhang. Computational self-assembly. *TCS*, to appear, 2007.

[63] A. Dal Palù, A. Dovier, and F. Fogolari. Constraint logic programming approach to protein structure prediction. *BMC Bioinformatics*, 5(186), 2004.

[64] V. Danos, J. Feret, W. Fontana, R. Harmer, and J. Krivine. Rule-based modelling of cellular signalling. In *Proceedings of CONCUR'07*, 2007.

[65] V. Danos and C. Laneve. Formal molecular biology. *Theor. Comput. Sci.*, 325(1):69–110, 2004.

[66] B. A. Davey and H. A. Priestley. *Introduction to Lattices and Order*. Cambridge University Press, 2nd edition, 2002.

[67] L. Davis. *Handbook of Genetic Algorithms*. Van Nostrand Reinhold, 1991.

[68] F. de Boer and C. Palamidessi. *Advances in logic programming theory*, chapter From Concurrent Logic Programming to Concurrent Constraint Programming, pages 55–113. Oxford University Press, 1994.

[69] F.S. de Boer, A. Di Pierro, and C. Palamidessi. Nondeterminism and infinite computations in constraint programming. *Theoretical Computer Science*, 151(1), 1995.

[70] H. De Jong. Modeling and simulation of genetic regulatory systems: A literature review. *Journal of Computational Biology*, 9(1):67–103, 2002.

[71] G. M. S. De Mori, C. Micheletti, and G. Colombo. All-atom folding simulations of the villin headpiece from stochastically selected coarse-grained structures. *Journal Of Physical Chemistry B*, 108(33):12267–12270, 2004.

[72] Luca Di Gaspero and Andrea Schaerf. EASYLOCAL++: An object-oriented framework for flexible design of local search algorithms. *Software — Practice & Experience*, 33(8):733–765, July 2003.

[73] A. Di Pierro and H. Wiklicky. A banach space based semantics for probabilistic concurrent constraint programming. In *Proceedings of CATS'98*, 1998.

[74] A. Di Pierro and H. Wiklicky. An operational semantics for probabilistic concurrent constraint programming. In *Proceedings of IEEE Computer Society International Conference on Computer Languages*, 1998.

[75] A. Di Pierro and H. Wiklicky. Probabilistic abstract interpretation and statistical testing. In H. Hermanns and R. Segala, editors, *Lecture Notes in Computer Science 2399*. Springer Verlag, 2002.

[76] A. Di Pierro, H. Wiklicky, and C. Hankin. Quantitative static analysis of distributed systems. *Journal of Functional Programming*, pages 1–43, 2005.

[77] K. A. Dill, S. Bromberg, K. Yue, K. M. Fiebig, D. P. Yee, P. D. Thomas, and H. S. Chan. Principles of protein folding - a perspective from simple exact models. *Protein Science*, 4:561–602, 1995.

[78] J. Ding and J. Hillston. On odes from pepa models. In *Proceedings of PASTA 2007*, 2007.

[79] M. Dorigo, G. Di Caro, and L. M. Gambardella. Ant algorithms for discrete optimization. Technical report, Universitè Libre de Bruxelles, 1999.

[80] M. Dorigo, V. Maniezzo, and A. Colorni. The ant system: Optimization by a colony of cooperating agents. *IEEE Transactions on Systems, Man, and Cybernetics -- Part B*, 26(1):29–41, 1996.

[81] G. Dueck and T. Scheuer. Threshold accepting: a general purpose optimization algorithm appearing superior to simulated annealing. *Journal of Computational Physics*, 90:161–175, 1990.

[82] H.-D. Ebbinghaus, J. Flum, and W. Thomas. *Mathematical Logic, second ed.* Springer-Verlag, 1994.

[83] L. Edelstein-Keshet. *Mathematical Models in Biology*. SIAM, 2005.

[84] R. Elber and M. Karplus. Multiple conformational states of proteins: a molecular dynamics analysis of myoglobin. *Science*, 235:318–321, 1987.

[85] M.B. Elowitz and S. Leibler. A synthetic oscillatory network of transcriptional regulators. *Nature*, 403:335–338, 2000.

[86] J.R. Faeder, M.L. Blinov, B. Goldstein, and W.S. Hlavacek. Rule-based modeling of biochemical networks. *Complexity*, 10:22–41, 2005.

[87] J. Felsenstein. Phylogeny inference package (version 3.2). *Cladistics*, 5:164–166, 1989. http://evolution.genetics.washington.edu/phylip.html.

[88] T.A. Feo and M.G.C. Resende. Greedy randomized adaptive search procedures. *Journal of Global Optimization*, 6:109–133, 1995.

[89] C. Floudas, J. L. Klepeis, and P. M. Pardalos. Global optimization approaches in protein folding and peptide docking. In F. Roberts, editor, *DIMACS Series in Discrete Mathematics and Theoretical Computer Science*. 1999.

[90] F. Fogolari, G. Esposito, P. Viglino, and S. Cattarinussi. Modeling of polypeptide chains as c-α chains, c-α chains with c-β, and c-α chains with ellipsoidal lateral chains. *Biophysical Journal*, 70:1183–1197, 1996.

[91] F. Fogolari, L. Pieri, A. Dovier, L. Bortolussi, G. Giugliarelli, A. Corazza, G. Esposito, and P. Viglino. A reduced representation of proteins: model and energy definition. *submitted to BMC Structural Biology*, 2006.

[92] W.J. Fokkink. *Introduction to Process Algebra*. EATCS. Springer, 2000.

[93] Swedish Institute for Computer Science. Sicstus prolog home page.

[94] M.R. Garey and D. S. Johnson. *Computers and Intractability: a Guide to the Therory of NP-Completeness*. Freeman, 1979.

[95] N. Geisweiller, J. Hillston, and M. Stenico. Relating continuous and discrete pepa models of signalling pathways. *Theoretical Computer Science*, 2008. in print.

[96] J. F. Gibrat, J. Garnier, and B. Robson. Further developments of protein secondary structure prediction using information theory. new parameters and consideration of residue pairs. *Journal of Molecular Biology*, 198(3):425–43, 1987.

[97] D. Gilbert and C. Palamidessi. Concurrent constraint programming with process mobility. In *Proceedings of CL 2000*, 2000.

[98] D. Gillespie. The chemical langevin equation. *Journal of Chemical Physics*, 113(1):297–306, 2000.

[99] D. Gillespie and L. Petzold. *System Modelling in Cellular Biology*, chapter Numerical Simulation for Biochemical Kinetics. MIT Press, 2006.

[100] D. T. Gillespie. A rigorous derivation of the chemical master equation. *Physica A*, 22:403–432, 1992.

[101] D.T. Gillespie. A general method for numerically simulating the stochastic time evolution of coupled chemical reactions. *J. of Computational Physics*, 22, 1976.

[102] D.T. Gillespie. Exact stochastic simulation of coupled chemical reactions. *J. of Physical Chemistry*, 81(25), 1977.

[103] K. Ginalski. Comparative modeling for protein structure prediction. *Curr. Opin. Struct. Biol.*, 16(2):172–177, 2006.

[104] F. Glover. Tabu search, part i. *ORSA journal of Computing*, 1:190–206, 1989.

[105] F. Glover. Tabu search, part ii. *ORSA journal of Computing*, 2:4–32, 1990.

[106] F. Glover and M. Laguna. *Tabu Search.* Kluwer Academic Publisher, 1997.

[107] F. Glover, E. Taillard, and D. de Werra. A users guide to tabu search. *Annals of Operational Research*, 41:3–28, 1993.

[108] A. Godzik, A. Kolinski, and J. Skolnick. Lattice representations of globular proteins: how good are they? *Journal of Computational Chemistry*, 14:1194–1202, 1993.

[109] A. Gosavi. *Simulation-Based Optimization.* Kluwer Academic Publisher, 2003.

[110] P.J.E. Goss and J. Peccoud. Quantitative modeling of stochastic systems in molecular biology by using stochastic petri nets. *Proc. Nat. Acad. Sci. USA*, 95:6750–6754, 1998.

[111] S. L. Le Grand and K. M. Merz jr. The application of the genetic algorithm to the minimization of potential energy functions. *J. Global Optim.*, 3:49–66, 1993.

[112] D. R. Greening. Parallel simulated annealing techniques. *Physica D*, 42:293–306, 1990.

[113] V. Gupta, R. Jagadeesan, , and V. A. Saraswat. Probabilistic concurrent constraint programming. In *Proceedings of CONCUR'97*, 1997.

[114] Vineet Gupta, Radha Jagadeesan, and Vijay A. Saraswat. Computing with continuous change. *Sci. Comput. Program.*, 30(1–2):3–49, 1998.

[115] J. Gutiérrez, J.A. Perez, C. Rueda, and F.D. Valencia. Timed concurrent constraint programming for analyzing biological systems. In *Proceedings of MeCBIC'06*, 2006.

[116] Rety. J. H. Distributed concurrent constraint programming. *Fundamentae Informaticae*, 34(3):323–346, 1998.

[117] P. J. Haas. *Stochastic Petri Nets.* Springer Verlag, 2002.

[118] S. Hardy and P. N. Robillard. Modeling and simulation of molecular biology systems using petri nets: Modeling goals of various approaches. *Journal of Bioinformatics and Computational Biology*, 2(4):619–637, 2004.

[119] S. P. Hastings and J. D. Murray. The existence of oscillatory solutions in the field-noyes model for the belousov-zhabotinskii reaction. *SIAM Journal on Applied Mathematics*, 28(3):678–688, 1975.

[120] R. A. Hayden and J. T. Bradley. Fluid-flow solutions in PEPA to the state space explosion problem. In *Proceedings of PASTA 2007*, 2007.

[121] S. J. Heims. *John von Neumann and Norbert Wiener: From Mathematics to the Technologies of Life and Death.* 3. Aufl., Cambridge, 1980.

[122] L. Henkin, J.D. Monk, and A. Tarski. *Cylindric Algebras, Part I.* North-Holland, Amsterdam, 1971.

[123] M. Hennessy. *Algebraic Theory of Processes.* The MIT Press, 1988.

[124] T. A. Henzinger. The theory of hybrid automata. In *LICS '96: Proceedings of the 11th Annual IEEE Symposium on Logic in Computer Science*, 1996.

[125] Holger Hermanns, Ulrich Herzog, and Joost-Pieter Katoen. Process algebra for performance evaluation. *Theor. Comput. Sci.*, 274(1-2):43–87, 2002.

[126] F. S. Hillier and G. J. Lieberman. *Introduction To Operations Research*. McGraw-Hill, 1995.

[127] J. Hillston. PEPA: Performance enhanced process algebra. Technical report, University of Edimburgh, 1993.

[128] J. Hillston. *A Compositional Approach to Performance Modelling*. Cambridge University Press, 1996.

[129] J. Hillston. Fluid flow approximation of PEPA models. In *Proceedings of the Second International Conference on the Quantitative Evaluation of Systems (QEST05)*, 2005.

[130] C.A.R. Hoare. *Communicating Sequential Processes*. Prentice Hall, 1985.

[131] J. Holland. Genetic algorithms and the optimal allocation of trials. *SIAM J. Computing*, 2:88–105, 1973.

[132] J. D. Honeycutt and D. Thirumalai. The nature of folded states of globular proteins. *Biopolymers*, 32:695–709, 1992.

[133] C.F. Huang and J.T. Ferrell. Ultrasensitivity in the mitogen-activated protein kinase cascade. *PNAS, Biochemistry*, 151:10078–10083, 1996.

[134] Joxan Jaffar and Michael J. Maher. Constraint logic programming: A survey. *Journal of Logic Programming*, 1994.

[135] Joxan Jaffar, Michael J. Maher, Kim Marriott, and Peter J. Stuckey. The semantics of constraint logic programs. *Journal of Logic Programming*, 37(1–3):1–46, 1998.

[136] D. T. Jones. Predicting novel protein folds by using fragfold. *Proteins*, 5S:127–131, 2001.

[137] N. G. Van Kampen. *Stochastic Processes in Physics and Chemistry*. Elsevier, 1992.

[138] H. Kawai, T. Kikuchi, and Y. Okamoto. A prediction of tertiary structures of peptide by the monte carlo simulated annealing method. *Protein Eng.*, 3:85–94, 1989.

[139] J. Kennedy and R. Eberhart. Particle swarm optimization. In *Proceedings of IEEE Internation Conference of Neural Networks*, 1995.

[140] S. Kirkpatrick, C. D. Geddat Jr., and M. P. Vecchi. Optimization by simulated annealing. *Science*, (220):671–680, 1983.

[141] H. Kitano. *Foundations of Systems Biology*. MIT Press, 2001.

[142] H. Kitano. Computational systems biology. *Nature*, 420:206–210, 2002.

[143] P.E. Kloeden and E. Platen. *Numerical solution of stochastic differential equations*. Springer-Verlag, New York, 1992.

[144] K. W. Kohn. Functional capabilities of molecular network components controlling the mammalian g1/s cell cycle phase transition. *Oncogene*, 16:1065–1075, 1998.

[145] K. W. Kohn. Molecular interaction map of the mammalian cell cycle control and dna repair systems. *Molecular Biology of the Cell*, 10:2703–2734, August 1999.

[146] K. W. Kohn, M. I. Aladjem, S. Kim, J. N. Weinstein, and Y. Pommier. Depicting combinatorial complexity with the molecular interaction map notation. *Molecular Systems Biology*, 2(51), 2006.

[147] K. W. Kohn, M. I. Aladjem, J. N. Weinstein, and Y. Pommier. Molecular interaction maps of bioregulatory networks: A general rubric for systems biology. *Molecular Biology of the Cell*, 17(1):1–13, 2006.

[148] A. Kolinski, A. Godzik, and J. Skolnick. A general method for the prediction of the three dimensional structure and folding pathway of globular proteins: application to designed helical proteins. *J. Chem. Phys.*, 98:7420–7433, 1993.

[149] J. Kostrowicki, L. Piela, B. J. Cherayil, and H. Scheraga. Performance of the diffusion equation method in searches for optimum structures of clusters of lennardjones atoms. *J. Phys. Chem.*, 95:4113–4119, 1991.

[150] J. Kostrowicki and H. A. Scheraga. Application of the diffusion equation method for global optimization to oligopeptides. *J. Phys. Chem.*, 96:7442–7449, 1992.

[151] J. Kostrowicki and H. A. Scheraga. *Global Minimization of Nonconvex Energy Functions: Molecular Conformation and Protein Folding*, chapter Some approaches to the multipleminima problem in protein folding, pages 123–132. Amer. Math. Soc., 1996.

[152] B. Kuhlman, G. Dantas, G. C. Ireton, G. Varani, B. L. Stoddard, and D. Baker. Design of a novel globular protein fold with atomic-level accuracy. *Science*, 302:1364–1368, 2003.

[153] T. G. Kurtz. Solutions of ordinary differential equations as limits of pure jump markov processes. *Journal of Applied Probability*, 7:49–58, 1970.

[154] T. G. Kurtz. Limit theorems for sequences of jump markov processes approximating ordinary differential processes. *Journal of Applied Probability*, 8:244–356, 1971.

[155] M. Kwiatkowska, G. Norman, and D. Parker. Probabilistic symbolic model checking with prism: A hybrid approach. *International Journal on Software Tools for Technology Transfer*, 6(2):128–142, September 2004.

[156] Brim L., Gilbert D., Jacquet J., and Kretinsky M. Multi-agent systems as concurrent constraint processes. In *Proceedings of SOFSEM 2001*, 2001.

[157] J. Lee, H. A. Scheraga, and S. Rackovsky. New optimization method for conformational energy calculations on polypeptides: Conformational space annealing. *Journal of Computational Chemistry*, 18(9):1222–1232, 1997.

[158] C. Levinthal. Are there pathways to protein folding? *J. Chim. Phys.*, 65:44–45, 1968.

[159] A. Liwo, J. Lee, D.R. Ripoll, J. Pillardy, and H. A. Scheraga. Protein structure prediction by global optimization of a potential energy function. *Proceedings of the National Academy of Science USA*, 96:5482–5485, 1999.

[160] J. Ma and J. E. Straub. Simulated annealing using the classical density distribution. *J. Chem. Phys.*, 101:533–541, 1994.

[161] A. D. Jr. MacKerell et al. All-atom empirical potential for molecular modeling and dynamics studies of proteins. *J. Phys. Chem. B*, 102:3586–3616, 1998.

[162] W.G. Macready, A.G. Siapas, and S.A. Kauffman. Criticality and parallelism in combinatorial optimization. *Science*, 271:56–59, 1996.

[163] M. Mamei, F. Zambonelli, and L. Leonardi. A physically grounded approach to coordinate movements in a team. In *Proceedings of* ICDCS, 2002.

[164] K. Marriott and P. Stuckey. *Programming with Constraints*. The MIT Press, 1998.

[165] H. H. Mcadams and A. Arkin. Stochastic mechanisms in gene expression. *PNAS USA*, 94:814–819, 1997.

[166] J.A. McCammon and S.C. Harvey. *Dynamics of Proteins and Nucleic Acids*. Cambridge University Press, 1987.

[167] P.G. Mezey. Potential energy hypersurfaces. *Studies in Physical and Theoretical Chemistry*, 53, 1987.

[168] A. Migdalas and P.M. Pardalos, editors. *Parallel Computing in Optimization*. Kluwer Academic Publishers, 1997.

[169] M. Milano and A. Roli. Magma: A multiagent architecture for metaheuristics. *IEEE Trans. on Systems, Man and Cybernetics - Part B*, 34(2), 2004.

[170] R.T. Miller, D.T. Jones, and J.M. Thornton. Protein fold recognition by sequence threading: tools and assessment techniques. *FASEB J.*, 10:171–178, 1996.

[171] R. Milner. Elements of interaction: Turing award lecture. *Commun. ACM*, 36(1):78–89, 1993.

[172] R. Milner. *Communication and concurrency*. Prentice Hall International (UK) Ltd., Hertfordshire, UK, UK, 1995.

[173] R. Milner. Turing, computing and communication. *Recollection of Alan Turing*, 1997.

[174] R. Milner. *Communicating and mobile systems: the π-calculus*. Cambridge University Press, New York, NY, USA, 1999.

[175] S. Miyazawa and R. L. Jernigan. Residue-residue potentials with a favorable contact pair term and an unfavorable high packing density term, for simulation and threading. *Journal of Molecular Biology*, 256(3):623–644, 1996.

[176] R. Motwani and P. Raghavan. *Randomized Algorithms*. Cambridge University Press, New York (NY), 1995.

[177] J. Moult. A decade of CASP: progress, bottlenecks and prognosis in protein structure prediction. *Curr. Op. Struct. Biol.*, 15:285–289, 2005.

[178] G. L. Nemhauser and L. A. Wolsey. Integer programming. In *Handbooks in Operations Research and Management Science*, chapter VI. North Holland, Amsterdam, 1989.

[179] A. Neumaier. Molecular modeling of proteins and mathematical prediction of protein structure. *SIAM Review*, 39:407–460, 1997.

[180] A. Neumaier. Complete search in continuous global optmization and constraint satisfaction. In A. Iserles, editor, *Acta Numerica 2004*. Cambridge University Press, 2004.

[181] J. R. Norris. *Markov Chains*. Cambridge University Press, 1997.

[182] R. M. Noyes and R. J. Field. Oscillatory chemical reactions. *Annual Review of Physical Chemistry*, 25:95–119, 1974.

[183] T. J. Oldfield and R. E. Hubbard. Analysis of c alpha geometry in protein structures. *Proteins*, 18(4):324–337, 1994.

[184] I. H. L. Osman and J. P. Kelly, editors. *Meta Heuristics, Theory and Applications*. Kluwer Academic Publisher, 1996.

[185] J.Y. Suh P. Jog and D. Van Gucht. Parallel genetic algorithms applied to the traveling salesman problem. *SIAM Journal of Optimization*, 1:515–529, 1991.

[186] Catuscia Palamidessi and Frank Valencia. A temporal concurrent constraint programming calculus. In Toby Walsh, editor, *Proc. of the 7th International Conference on Principles and Practice of Constraint Programming*, volume 2239, pages 302–316. LNCS, Springer-Verlag, 2001.

[187] P. M. Pardalos, L. S. Pitsoulis, T. Mavridou, and Resende M.G.C. Parallel search for combinatorial optimization: Genetic algorithms, simulated annealing and grasp. In A. Ferreira and J. Rolim, editors, *Proceedings of Irregular'95, Lecture Notes in Computer Science*, volume 980, pages 317–331. Springer-Verlag, 1995.

[188] P.M. Pardalos, G. Xue, and P.D. Panagiotopulos. Parallel algorithms for global optimization. In *Solving Combinatorial Optimization Problems in Parallel*, pages 232–247, 1996.

[189] B. Park and M. Levitt. Energy functions that discriminate x-ray and near-native folds from well-constructed decoy. *Proteins: Structure Function and Genetics*, 258:367–392, 1996.

[190] James Lyle Peterson. *Petri Net Theory and the Modeling of Systems*. Prentice Hall PTR, 1981.

[191] J. W. Pinney, D. R. Westhead, and G. A. McConkey. Petri net representation in systems biology. *Biochemical Society Transactions*, 31(6):1513–1515, 2003.

[192] G.D. Plotkin. A structural approach to operational semantics. *J. Log. Algebr. Program.*, 60-61:17–139, 2004.

[193] W.H. Press, S.A. Teukolsky, W.T. Vetterling, and B.P. Flannery. *Numerical Recipes in C++ : The Art of Scientific Computing*. Cambridge University Press, 2002.

[194] C. Priami. Stochastic π-calculus. *The Computer Journal*, 38(6):578–589, 1995.

[195] C. Priami and P. Quaglia. Modelling the dynamics of biosystems. *Briefings in Bioinformatics*, 5(3):259–269, 2004.

[196] C. Priami and P. Quaglia. Beta binders for biological interactions. In *Proceedings of Computational methods in system biology (CMSB04)*, 2005.

[197] C. Priami, A. Regev, E. Y. Shapiro, and W. Silverman. Application of a stochastic name-passing calculus to representation and simulation of molecular processes. *Inf. Process. Lett.*, 80(1):25–31, 2001.

[198] J. Puchinger and G. R. Raidl. Combining metaheuristics and exact algorithms in combinatorial optimization: A survey and classification. In Springer-Verlag, editor, *Proceedings of the First International Work-Conference on the Interplay Between Natural and Artificial Computation*, volume 3562 of *LNCS*, pages 41–53, 2005.

[199] D. Qiu, P. Shenkin, F. Hollinger, and W. Still. The gb/sa continuum model for solvation. a fast analytical method for the calculation of approximate born radii. *J. Phys. Chem.*, 101:3005–3014, 1997.

[200] S. Ramsey, D. Orrell, and H. Bolouri. Dizzy: stochastic simulation of large-scale genetic regulatory networks. *Journal of Bioinformatics and Computational Biology*, 3(2):415–436, 2005.

[201] C. V. Rao and A. P. Arkin. Stochastic chemical kinetics and the quasi-steady state assumption: Application to the gillespie algorithm. *Journal of Chemical Physics*, 118(11):4999–5010, March 2003.

[202] V. J. Rayward-Smith, I.H. Osman, C. R. Reeves, and G.D. Smith. *Modern Heuristics Search Methods*. John Wiley and Sons, 1996.

[203] A. Regev and E. Shapiro. Cellular abstractions: Cells as computation. *Nature*, 419, 2002.

[204] M. Resende, P. Pardalos, and S. Duni Ekşioğlu. Parallel metaheuristics for combinatorial optimization. In R. Correa et al., editors, *Models for Parallel and Distributed Computation - Theory, Algorithmic Techniques and Applications*, pages 179–206. Kluwer Academic, 2002.

[205] W. G. Richards. Calculation of conformational free energy of histamine. *J. Theor. Biol.*, 43:389, 1974.

[206] B. D. Ripley. *Stochastic Simulation*. Wiley, New York, 1987.

[207] A. Romanel, L. Demattè, and C. Priami. The beta workbench. Technical Report TR-03-2007, Center for Computational and Systems Biology, Trento, 2007.

[208] B. Rost. Protein Secondary Structure Prediction Continues to Rise. *J. Struct. Biol.*, 134:204–218, 2001.

[209] B. Rost, R. Schneider, and C. Sander. Protein fold recognition by prediction-based threading. *J. Mol. Biol.*, 1996.

[210] W. Rudin. *Principles of Mathematical Analysis*. McGraw-Hill, 1976.

[211] J. Rutten, M. Kwiatkowska, G. Norman, and D. Parker. *Mathematical Techniques for Analyzing Concurrent and Probabilistic Systems*, volume 23 of *CRM Monograph Series*. American Mathematical Society, 2004.

[212] D. Sangiorgi and D. Walker. *The π-calculus: a theory of mobile processes*. Cambridge University Press, 2001.

[213] V. Saraswat and M. Rinard. Concurrent constraint programming. In *Proceedings of 18th Symposium on Principles Of Programming Languages (POPL)*, 1990.

[214] V. A. Saraswat. *Concurrent Constraint Programming*. MIT press, 1993.

[215] V. A. Saraswat, M. Rinard, and P. Panangaden. Semantics foundations of concurrent constraint programming. In *Proceedings of POPL*, 1991.

[216] M. Schwehm. Parallell stochastic simulation of whole cell models. In *Proceeding of ICSB 2001*, 2001.

[217] D. Scott. Domains for denotational semantics. In *Proceedings of ICALP*, 1982.

[218] E. Seneta. *Non-Negative Matrices and Markov Chains*. Springer-Verlag, 1981.

[219] P. Serafini. *Ottimizzazione*. Zanichelli, 2000.

[220] B. E. Shapiro, A. Levchenko, E. M. Meyerowitz, Wold B. J., and E. D. Mjolsness. Cellerator: extending a computer algebra system to include biochemical arrows for signal transduction simulations. *Bioinformatics*, 19(5):677–678, 2003.

[221] Paul C. Shields. *The Ergodic Theory of Discrete Sample Paths*, volume 13 of *Graduate Studies in Mathematics*. American Mathematical Society, Providence, Rhode Island, 1996.

[222] J. K. Shin and M. S. Jhon. High directional monte carlo procedure coupled with the temperature heating and annealing method to obtain the global energy minimum structure of polypeptides and proteins. *Biopolymers*, 31:177–185, 1991.

[223] J. Skolnick and A. Kolinski. Reduced models of proteins and their applications. *Polymer*, 45:511–524, 2004.

[224] Z. Slanina. Does the global energy minimum always also mean the thermodynamically most stable structure? *J. Mol. Str.*, 206:143–151, 1990.

[225] F. H. Stillinger and T. A. Weber. Nonlinear optimization simplified by hypersurface deformation. *J. Stat. Phys.*, 52:1492–1445, 1988.

[226] J. E. Straub. *New Developments in Theoretical Studies of Proteins*, chapter Optimization techniques with applications to proteins. World Scientific, 1997.

[227] S. H. Strogatz. *Non-Linear Dynamics and Chaos, with Applications to Physics, Biology, Chemistry and Engeneering*. Perseus books, 1994.

[228] D'Arcy Wentworth Thompson. *On Growth and Form.* ., 1917.

[229] T. Tian and K. Burrage. Parallel implementation of stochastic simulation for large-scale cellular processes. In *Proceeding of Eighth International Conference on High-Performance Computing in Asia-Pacific Region (HPCASIA'05)*, pages 621–626, 2005.

[230] T. Toma and S. Toma. Folding simulation of protein models on the structure-based cubooctahedral lattice with the contact interactions algorithm. *Protein Science*, 8:196–202, 1999.

[231] A. Šali, E. Shakhnovich, and M. Karplus. How does a protein fold? *Nature*, 369:248–251, 1994.

[232] T. Veitshans, D. Klimov, and D. Thirumalai. Protein folding kinetics: timescales, pathways and energy landscapes in terms of sequence-dependent properties. *Folding & Design*, 2:1–22, 1996.

[233] M.G.A. Verhoeven and E.H.L. Aarts. Parallel local search. *Journal of Heuristics*, 1:43–65, 1995.

[234] J. M. G. Vilar, H. Yuan Kueh, N. Barkai, and S. Leibler. Mechanisms of noise resistance in genetic oscillators. *PNAS*, 99(9):5991, 2002.

[235] A. Villanueva. Model Checking for the Concurrent Constraint Paradigm (Abstract of Ph.D. Thesis). *Bulletin of the European Association for Theoretical Computer Science (EATCS)*, (82):389–392, February 2004.

[236] E. O. Voit. *Computational Analysis of Biochemical Systems*. Cambridge University Press, 2000.

[237] L. von Bertalanffy. *General System theory: Foundations, Development, Applications.* George Braziller, New York, 1968.

[238] N. Wiener. *Cybernetics or Control and Communication in the Animal and the Machine.* MIT Press, Cambridge, MA, 1948.

[239] D. J. Wilkinson. *Stochastic Modelling for Systems Biology.* Chapman & Hall, 2006.

[240] L. T. Wille and J. Vennik. Computational complexity of the ground-state determination of atomic clusters. *Journal of Physics A: Mathematical and General,* 18(8):L419–L422, 1985.

[241] M. Yagiura and T. Ibaraki. On metaheuristic algorithms for combinatorial optimization problems. *Systems and Computers in Japan,* 32(3):33–55, 2001.

[242] Y. Zhang, A. K. Akaraki, and J. Skolnick. Tasser: an automated nethod for the prediction of protein tertiary structures in casp6. *Proteins,* 7S:91–98, 2005.